D0215030

STALIN

AND THE SOVIET SCIENCE WARS

STALIN

AND THE SOVIET SCIENCE WARS

ETHAN POLLOCK

PRINCETON UNIVERSITY PRESS - PRINCETON AND OXFORD

© 2006 by Princeton University Press
Published by Princeton University Press, 41 William Street, Princeton,
New Jersey 08540
In the United Kingdom: Princeton University Press, 3 Market Place,
Woodstock, Oxfordshire OX20 1SY

Library of Congress Cataloging-in-Publication Data
Pollock, Ethan, 1969–
Stalin and the Soviet Science Wars / Ethan Pollock.
p.   cm.
Includes bibliographical references and index.
ISBN-13: 978-0-691-12467-4 (cloth: alk. paper)
ISBN-10: 0-691-12467-1 (cloth: alk. paper)
1. Communism and science—Soviet Union. 2. Science and state—Soviet
Union. 3. Soviet Union—Politics and government—1936–1953. 4. Stalin,
Joseph, 1879–1953. I. Title

HX541.P63 2006
338.947′0609044—dc22   2005035125

British Library Cataloging-in-Publication Data is available

This book has been composed in Goudy

Printed on acid-free paper.∞

pup.princeton.edu

Printed in the United States of America

10 9 8 7 6 5 4 3 2 1

FOR MY PARENTS
AND IN MEMORY OF REGGIE AND ALEX

CONTENTS

CHAPTER 7
"Everyone Is Waiting": Stalin and the Economic Problems

CHAPTER 8

FIGURES

STALIN

AND THE SOVIET SCIENCE WARS

CHAPTER 1

INTRODUCTION

Stalin, Science, and Politics after the Second World War

Joseph Stalin collected many encomiums while ruling the Soviet Union. At various times the Soviet press called him, among other things: "the Standard-bearer of Peace," "the Great Helmsman of the Revolution," "the Leader of the International Proletariat," "Generalissimo," and "the Father of Nations." In the years following the Second World War he assumed yet another title: "the coryphaeus of science."[1] As the "leader of the chorus"—or coryphaeus—Stalin stood on the podium while Soviet scientists sang in rhythm to the commanding movements of his baton.

Stalin tried to live up to the ideal of a man who united political power and intellectual acumen. Between the end of the Second World War and his death in 1953 he intervened in scientific debates in fields ranging from philosophy to physics.[2] In late 1946, when Stalin was sixty-seven years old and exhausted from the war, he schooled the USSR's most prominent philosopher on Hegel's role in the history of Marxism. In 1948, while the Berlin crisis threatened an irreparable rift between the United States and the USSR, Stalin wrote memos, held meetings, and offered editorial comments in order to support attacks against Mendelian genetics. In 1949, with the first Soviet atomic bomb test only months away, Stalin called off an effort to purge Soviet physics of "bourgeois" quantum mechanics and relativity. In the first half of 1950 he negotiated a pact with the People's Republic of China and discussed plans with Kim Il Sung about invading South Korea, while also writing a combative article on linguistics, carefully orchestrating a coup in Soviet physiology, and meeting with economists three times to discuss a textbook on political economy. In some cases he denounced whole fields of scholarship, leading to the firing and occasional arrest of their proponents. His efforts to unmask errors in science were paralleled by an equally intense drive to show how each discipline could contribute to building communism and serve as a symbolic weapon of Soviet superiority in the battle with the West along an "ideological front."

Why was Stalin so keen to be a scholar? His direct involvement in academic disputes reveals a side of the aging dictator that supplements what we have long known about him from the extensive memoir literature. He took ideology seriously. He was not simply a megalomaniac and reclusive old man who used scholarly debates only to settle political problems. (After all, he had much more direct ways of taking care of things he did not like.) The evidence shows he was far more concerned about ideas than was previously known. We do not have to accept the intellectual value of Stalin's proclamations about biology, linguistics, physiology, or political economy to recognize that he consistently spent time on the details of scholarly disputes.

Applying Marxist-Leninist principles to academic controversies often led to unpredictable results. Even those members of Stalin's inner circle who were responsible for ideology had to wait for word from the coryphaeus before they could be confident that they understood the outcomes he had in mind. For his part, Stalin's strategies for solving scholarly conflicts evolved in response to ideas put forth by scientists themselves. When he did reveal his judgments, others were left with the unenviable task of interpreting his words and working out their implications for a wide range of fields. In this sense Marxist-Leninist ideology was often subject to reformulation.

This book analyzes the content of Stalin's scientific forays, places them within the context of the broader academic disputes, and then traces their impact on both domestic high politics and the Soviet conceptualizations of the Cold War. In order to do this, the story moves up and down the Soviet system, from the institutes and universities where scientific debates often began and where their effects became apparent, to the presidium of the colossal Academy of Sciences, to deliberations in the Secretariat of the Central Committee, and finally to Stalin's office and desk, where the leader passed final judgments.

Controversies erupted in many academic fields in the 1940s and early 1950s. Six stand out because of their broad implications and because Stalin and his closest lieutenants in the Soviet government and the Communist Party directly intervened in them. These six debates—or "discussions," as they were often called—took place in philosophy, biology, physics, linguistics, physiology, and political economy.[3] Stalin's active participation in these debates demonstrates that more was at stake than scholarly disagreements: the science wars of the late Stalin period encompassed themes crucial to the Party's legitimacy and fundamental to the Soviet worldview in the early Cold War. Marxist-Leninist "scientific philosophy" provided the foundation for the ideology that underpinned the state and society. Physiology and biology had a direct bearing on the new "Soviet man" that the system tried to create and on nature, which communism promised to transform. Quantum mechanics and relativity in physics potentially challenged Marxist-Leninist materialist epistemology even as they seemed crucial for the development of atomic weapons. Linguistics encompassed issues of consciousness, class, and nationality. And political economy required a critique of capitalism, a justification of Soviet socialism, and a road map for achieving communism in the USSR and throughout the world.[4]

Stalin did not venture into scientific laboratories, conduct specific experiments, or solve equations. Yet he insisted that science was intertwined with the foundations of socialism and with the Party's raison d'être. Thousands of newly accessible and previously unexplored documents from Communist Party, Russian State, and Academy of Sciences archives reveal that he was determined—at times even desperate—to show the scientific basis of Soviet Marxism. As both an editor and an author, Stalin actively engaged with the content of scholarly work and contemplated its overall implications for Marxism-Leninism. His memos and top secret documents are saturated with the same Marxist-Leninist language, categories, and frames for understanding the world that appeared in the public discourse. He did not keep two sets of books, at least on ideological questions.[5]

Under Stalin's guidance, the USSR went further than any previous state in placing the support of science at the center of its stated purpose.[6] As a Marxist presiding over an agrarian country, Stalin was eager to modernize as quickly as possible. He believed that science provided the key to updating and industrializing the economy. Principles of scientific management would improve not only industrial production but all other aspects of societal development. Like Engels and Lenin before him, Stalin understood Marxism as a science inextricably tied to the methodology and laws of the natural sciences. Marxism-Leninism claimed to provide a "science of society" that would help to create a "kingdom of freedom" on earth. The Party's political authority relied on the perceived rationality and scientific basis of its actions. If Marxism-Leninism was scientific, and science would flourish if it was based on Marxist principles, it followed that science and Soviet Marxism should mutually reinforce each other. They led to the same discoveries about the nature of things and, together, progressed steadily to absolute truths.

Science played a unique role in Soviet ideology. When Soviet citizens publicly spoke or wrote about Soviet ideology, they were referring to a set of ideas identified and propagated by the regime and used to justify the superiority of the Soviet state. In principle, these ideas were derived from interpretations of canonical texts by Marx, Engels, and Lenin and were supposed to reflect and shape Soviet reality. They were supposed to be all-encompassing and internally consistent with "Party lines" defining the parameters of acceptable positions within various fields of thought. Soviet ideology contrasted with "bourgeois ideology"—a pejorative term depicting ideas in the Western political "superstructure" that reflected the capitalist "economic base." By definition, Soviet ideology was an accurate depiction of the material world, while bourgeois ideology consisted of lies and illusions that helped the capitalists to maintain power. The regime strictly upheld its prerogative to judge every activity on ideological grounds. But what about the cases when science and Soviet ideology seemed to contradict one another? Unlike the literary or artistic intelligentsia, whose challenges to the Party's authority were based on subjective notions of justice and moral truth that the Party could simply reject, scientists based their autonomy on very limited fields of expertise that provided them with specific access to objective laws. Scientists claimed that their work reflected reality, just like Soviet ideology.

The relationship between science and the Party evolved over the course of Soviet rule. During the 1920s the sciences, particularly the natural sciences, were relatively free from a radical Bolshevik agenda that sought to revolutionize thought in the name of building proletarian culture.[7] While theorists debated the meaning of dialectical materialism as a Marxist philosophy of science, Lenin defended "bourgeois technical experts" and the contribution they could make to modernizing the state. The regime denounced bourgeois literature, art, social policies, and the like, but it supported bourgeois scientists. During the Great Break of the late 1920s and early 1930s, however, zealous Marxist-Leninist philosophers promoted some scientific theories as "proletarian" and rejected others as "bourgeois." In an attempt to create "red specialists," activists and young students pushed to expose, fire, and arrest so-called saboteurs and wreckers among the "bourgeois experts."[8] In 1931 Stalin called for an end to the radical upheaval of the period, and subsequently the Party supported a calcified dialectical materialism based more on loyalty to the Party than on specific philosophical tenets. Scientists and the regime reached a new modus vivendi in which the Party supported scientific research while retaining control over scientific planning.[9] By the end of the 1930s young scientists who owed their education to progressive Soviet policies tended to be more sympathetic to Marxism-Leninism, and young leaders in the Party and state apparatus who had received training in technical disciplines tended to see themselves as part of a new Soviet intelligentsia.[10]

The Second World War altered the relationship between ideology and science in three crucial ways. First, scientists found themselves relatively free from Party oversight. Second, the atmosphere of international cooperation exemplified by the antifascist Grand Alliance created an opportunity for Soviet scholars to participate in "world science" and weakened the distinctions between "bourgeois" and "proletarian" science. And third, the development of atomic weapons, radar, and antibiotics during the war clarified that science was a crucial component of national security, which increased Party support and scrutiny. In these fields, science in the West was in the lead and could not be dismissed. The wartime mood was summed up at an international celebration of the 220th anniversary of the Academy of Sciences in the Kremlin in June 1945. With Stalin and foreign scientists in attendance, the Soviet minister of foreign affairs, Viacheslav Molotov, proposed a toast for "the development of close collaboration between Soviet and world science."[11]

The opportunity for cooperation in science did not last long. In February 1946, Stalin delivered a speech blaming capitalist policies for the outbreak of the two world wars and outlining a plan to guarantee that the USSR would be militarily prepared for the next global conflict. From the perspective of American policy makers, the Cold War was under way.[12] The speech also assured that science would be an important sphere of international competition. "I have no doubt that if we give our scientists proper assistance," Stalin said, "they will be able in the very near future not only to overtake but even outstrip the achievements of science beyond the borders of our country."[13] The Cold War was not just about geopolitics and military conflicts. It also pitted two ways of organizing science against one another.

Stalin provided practical support for the effort to surpass foreign science. In early 1946 Stalin told Igor Kurchatov, the physicist in charge of the Soviet atomic bomb project, "our state has suffered much, yet it is surely possible to ensure that several thousand people can live very well, and several thousand people better than very well, with their own dachas, so that they can relax, and with their own cars."[14] This was true not only of physicists working to end the American atomic monopoly. The rising tide raised all ships: funding for the Academy of Sciences expanded rapidly, as did the number of institutes and the number of scholars working in them. In turn for their loyalty and hard work, Stalin gave scientists material comforts that were extremely rare in the USSR at the time.[15]

Science became a sphere of Cold War competition in ways that went beyond national security. Stalin assigned Soviet scholars two key roles on the "ideological front" of the Cold War: they had to criticize Western ideas, and they had to export Soviet ideas to newly emerging socialist states in Eastern Europe and Asia. Sustaining the argument that communism was the only viable way to organize society required a certain ideological coherency, which scholars could provide. One of the best ways to prove the merits of a materialist worldview was to show that adhering to it inevitably led to scientific breakthroughs. Soviet intellectual achievements could serve as symbolic measures of the superiority of the Soviet system. Scholars from every discipline joined the battle along the ideological front. Stalin implored one group of economists, for instance, to recognize the broader significance of their work, which would be "read by Americans and Chinese . . . studied in all countries. . . . It will be a model for everyone."[16] Soviet scholars had to espouse universal theories in an effort to win the hearts and minds of people around the globe. Stalin saw the real need for and value of science, hence his own involvement.

Despite the value of scholars in Cold War competition, Stalin never fully trusted their loyalty. The lingering appeal of international cooperation and "world science" challenged the strict dichotomy between East and West that the Party emphasized. Even the stunning success of the USSR in the Second World War, an apparent vindication of Stalin's policies, exacerbated tensions between the regime and the scholarly elite. Soviet citizens hoped that victory in war would bring improvements in living standards and increased ideological flexibility. Instead, financial instability, widespread famine, severe health care problems, and the Party's attempt to gear the economy for the Cold War led to unexpected sacrifice by ordinary citizens.[17] Stalin believed that this social dissatisfaction could undermine confidence in the system more generally. So, rather than loosening its grip, the Party tightened it and looked for scapegoats who could be blamed for the persistent hardships. Soviet intellectuals, including scientists, who had actively developed contacts with foreigners during the relative openness of the wartime alliance were easy targets.

As international tensions rose, Stalin moved systematically to reestablish control over all sectors of society. Scientific discussions became a means by which the Party could ensure scientists' loyalty to the state and to Party principles. Stalin worried that Soviet intellectuals had fallen under the influence of Western culture. In 1946, at his boss's behest, Party secretary Andrei Zhdanov led an attack against Soviet writers for their "formalism" and "subservience to bourgeois culture." Similar denunciations followed in music and art in a campaign that became known as the zhdanovshchina. These internal struggles were clearly connected to the international situation. In 1947 Stalin ordered Zhdanov to deliver a major policy speech declaring that the world was divided into "two camps" and that there could be no neutral parties between them.[18] Like everyone else, scientists had to conform to the bipolarity of the Cold War. In 1947 Stalin told the popular writer Konstantin Simonov, "if you take our intelligentsia, scientific intelligentsia, professors, physicians—they are not sufficiently inculcated with the feeling of Soviet patriotism. They have unjustified admiration for foreign culture."[19] Soon afterward the Central Committee distributed a closed letter to all Party members condemning "servility to the West" and calling on the intelligentsia to "defend the interest and honor of the Soviet state."[20]

Some scholars responded to the Party's call to arms by reviving in their own disciplines the class categories and divisions from the debates of the late 1920s and early 1930s. They accused some Soviet scientists of "bourgeois" values and discredited their ideas as manifestations of "bourgeois" science. With the world divided into "two camps," the Party demanded that Soviet science contribute to the advancement of socialism and exemplify the superiority of socialist ideology. The invocation of the vocabulary of class warfare was at odds, however, with a more recent drive to praise all things Russian. Beginning in the mid to late 1930s—that is, during the retreat from the radicalism of the Great Break—the Party began to cite the positive attributes of Russians and Russian traditions as a means of explaining the special role of the USSR in the world. The Second World War or "Great Patriotic War," as it was known in the USSR, only strengthened this shift.[21] In this nationalist vein, it was claimed that Russians had laid the foundation for the natural sciences, had invented the radio and airplane, and were responsible for many of the greatest ideas the world had ever known. Postwar ideology required scientists in every field to work out a new set of tenets that encompassed the seemingly contradictory elements of class and Russocentrism. The new Soviet patriotism—in its Russocentric manifestation—became a standard for judging the value of scientists if not their science.

The opposite of patriotism was subservience to the West. By the late 1940s the struggle to ensure loyalty among Soviet citizens had evolved into an effort to purge Soviet society of all "cosmopolitan" influences. Officially, the "anticosmopolitan" campaign targeted anyone with foreign contacts and those who had ever expressed admiration for foreign culture. In practice, cosmopolitanism quickly became associated with Jews. Local organizations responded by firing thousands of Jews because of their alleged disloyalty. The Party arrested many prominent Jews, sentencing them to death or years in forced labor camps. Because of their disproportionate representation in the academy, Jews in scientific fields were particularly vulnerable to the campaign.[22] Secret memos show how troubled Party leaders became when they realized that Soviet physics, economics, and other fields were dominated by Jews and other ethnic minorities. The xenophobia of "anticosmopolitanism" permeated the scientific discussions. Rather than simply determining whether a scientific theory corresponded to the latest interpretation of Marx and Lenin's writings, meetings became forums for denouncing individuals—almost always non-Russians—for maintaining contacts with and citing foreign scientists.

With so much at stake, it is little wonder that the Soviet Union's most powerful Party and government leaders—including Zhdanov, Georgii Malenkov, and Lavrenty Beria—got involved in scientific discussions. Scientific disputes became particularly heated in part because they fell under the jurisdiction of both the Party and the state and as such became focal points for clashes between Stalin's lieutenants. Andrei Zhdanov's power derived from his position in the Party, where he was in charge of defining and enforcing unanimity in Soviet ideology and culture. Scientific controversies left him vulnerable because they revealed potential doctrinal fault lines. In contrast, his rivals Malenkov and Beria derived their strength from their dominance of the state apparatus, including ministries that funded and monitored science and education. They could use scientific discussions to enhance their own power by highlighting Zhdanov's inability to solve the persistent problems along the ideological front. Beyond that, doctrinal issues interested them primarily as a means to gain favor with Stalin.

The six postwar scientific meetings addressed a common theme: in each case Party leaders and scholars struggled to make space for both Soviet ideology and Soviet science. Each discussion began with scholarly disagreements in scientific institutes, in popular and scientific publications, and in the Central Committee. Scientific administrators such as Sergei Vavilov, the president of the Academy of Sciences, and Sergei Kaftanov, the minister of education of the USSR, actively oversaw disputes and forwarded their opinions to Stalin and other Party secretaries. Individual scientists presented their arguments to the Party as well, either by publishing articles or by appealing directly to patrons in the highest echelons of the Party. In the Central Committee, responsibility for monitoring scholarship rested with the Agitation and Propaganda Administration (Agitprop) and within it the Science Section.[23] Because Party personnel at this level did not have the authority to settle major conflicts on their own, particularly complicated or troublesome disputes made their way up the Party apparatus to the Party Secretariat.

In the second stage of each discussion, Party leaders and scholars set about settling the scientific conflicts and defining a unified ideological position. As controversies became more heated, Party secretaries reviewed the analyses and plans of their subordinates. Depending on the nature and seriousness of the matter, decisions would either be made by the Secretariat or passed along to the apex of Party power, Stalin and the Politburo. At times Stalin dramatically reversed decisions made at lower levels. The threat of such actions by Stalin left Party organizers and scholars alike in a state of constant uncertainty about the validity and proper meaning of their carefully crafted recommendations.

The decisive meeting in each field was organized to strike the proper balance between the Party's role in determining the outcomes of debates and the importance of scholarly participation. In other spheres of Soviet life the Party did not hesitate to use decrees, speeches, and publications to articulate and uphold ideological tenets. These techniques would not suffice when the goal was to reconcile Marxism-Leninism with major scientific findings. Instead, Stalin and the Central Committee insisted on the scientific discussions. Scholars, in the course of debates that were closely observed (but never totally controlled) by the Party, were supposed to forge an understanding of their disciplines that was in harmony with ideology, even when the Party's views were not clear to them, or indeed to the Party supervisors themselves.

In order to help formulate ideologically correct science, the Party often promoted what can be thought of as "comrade scientists"—that is, heroic figures combining both ideological vigor and scholarly expertise. The Nobel laureate physiologist Ivan Pavlov, for instance, was posthumously presented as a great scientist whose materialist philosophy and outstanding scientific advancements went hand in hand. The linguist Nikolai Marr was posthumously knocked off a similar pedestal, but only after scientific and Party administrators alike had spent years declaring that his theories had done more to advance Marxist linguistics than anything ever written. Georgii Aleksandrov in philosophy and Trofim Lysenko in biology also embodied a blend of scholarship and Party-mindedness, but for differing lengths of time and with strikingly different outcomes. Of course, Stalin was the ultimate comrade scientist. In all six scholarly discussions, Stalin either contributed an essay of his own or intervened indirectly through instructions to Party leaders or scientists. While he was alive, Stalin was the only person in the Soviet Union who, by definition, never erred on either ideological or scholarly issues. Indeed, his role was so important that major scientific discussions could not be settled until Stalin's views were known.

Stalin *ex machina* was decisive in principle. But confusion over the proper interpretation of the new Party line continued even after the discussions' official conclusions, leading most disciplines into long periods of stagnation. Efforts in the Academy of Sciences and the Ministry of Education to define and enforce a unified ideology based on Stalin's dictums proved futile, and Agitprop and the Science Section continued to lament what they deemed to be a crisis in Soviet ideology. Widely publicized declarations notwithstanding, inconsistencies abounded. The aftermaths of the discussions suggest that scientists, administrators, and Central Committee secretaries were all caught off guard by the direction of scholarly disputes. Outcomes from one discussion did not translate into clear lessons for other disciplines. Far from displaying a carefully formulated and executed message, each successive discussion revealed apparent contradictions in Soviet ideology that in turn resulted in further debate and floods of letters to the Central Committee demanding clarifications.

Although the debates shared certain structural features, the specifics of what was discussed and the conclusions they reached varied considerably. The discussion in philosophy, the subject of chapter 2, began in December 1946 in the Kremlin when Stalin informed an elite group of leaders and scholars that Aleksandrov's prizewinning history of Western European philosophy had overstated the influence of Hegel and other German philosophers on Marxism. Despite being head of Agitprop, Aleksandrov had misinterpreted what Stalin required of scholars working on the ideological front of the Cold War. The discussion culminated in June 1947 with a meeting at the Central Committee attended by a wide range of the Soviet political and scientific elite. Stalin maneuvered behind the scenes and Zhdanov, the Party's second in command, publicly attacked Aleksandrov and the discipline of philosophy in general.

A little over a year later, in the summer of 1948, Lysenko took advantage of his personal favor with Stalin to hold a meeting of the All-Union Agricultural Academy. As discussed in chapter 3, Lysenko revealed at the meeting that the

Party supported his outright suppression of Western, Mendelian genetics and favored a homegrown, Soviet theory that emphasized the inheritance of acquired characteristics. The story of Lysenko's monopolization of Soviet biology—what Stephen Jay Gould called "the most chilling passage in all the literature on twentieth-century science"—has dominated scholarship on Soviet science.[24] The context of the other scientific discussions clarifies that the situation in biology constituted only one part of a much broader effort to come up with a coherent understanding of the relationship between Soviet ideology and science.

The next meeting, discussed in chapter 4, had a very different outcome: it was canceled. The All-Union Physics Conference planned for early 1949, and modeled on the 1948 biology meeting, never took place, despite months of careful preparation by physicists and Party philosophers. A select number of physicists formed a cohesive and savvy group that managed to convince the conference organizers that the national meeting would never reach a consensus about what, exactly, ideologically correct physics would look like. Furthermore, Beria, the brutal police chief whom Stalin had put in charge of the Soviet atomic weapons project, recognized the expedience of protecting the scientists under his charge from attacks by ideological zealots. Physicists adeptly translated the importance of atomic weapons research into unprecedented control over their own profession. Andrei Sakharov, a young weapons designer at the time, participated in some of these political maneuverings and took away from them crucial lessons that he would later apply as a dissident.

In the spring and early summer of 1950, two more discussions—about linguistics and about physiology—took place, one right after the other. They are the subjects of chapters 5 and 6 respectively. In May and June 1950 *Pravda* printed dozens of conflicting articles on the state of Soviet linguistics. Then, shockingly, Stalin intervened with an essay overturning the previously held orthodoxy and suggesting that language was neither part of the economic base nor part of the political superstructure, two core categories of Marxist ideology. He also suggested that scientific innovation required free and open discussions. After the coryphaeus of science had spoken, scholars in every field, not just linguistics, scrambled to interpret the implications of the new pronouncements for their own work and for science more generally.

In late June and early July, within days of the conclusion of the linguistics discussion, hundreds of physiologists convened in Moscow at a meeting organized to ensure that Soviet physiology followed a rigid interpretation of Pavlov's work. With heavy-handed coaxing from Stalin and the Science Section, a number of prominent physiologists defended Pavlov's insistence that conditioned reflexes provided the keys to understanding complex behavior in all animals, including humans. The Politburo set out to enforce the meeting's conclusion and charged the Science Section with overseeing a scientific council that continued to repress those who defended a broader understanding of Pavlov's scientific contribution and legacy.

Finally, chapter 7 addresses a month-long meeting in late 1951 where hundreds of economists and political leaders gathered at the Central Committee to discuss a draft of a political-economy textbook. Stalin intended the book to be used in the Soviet Union and throughout the expanding socialist camp and therefore fretted over even the smallest details. Party Secretaries Malenkov and Suslov chaired the daily sessions, while Stalin took the lead role in organizing the discussion and shaping its outcome. In response to the meeting, he also published a long essay in which he declared that "Marxism regards laws of science—whether they are laws of natural science or laws of political economy—as the reflection of objective processes which take place independently of the will of man." The laws of science provided the standard by which to judge the validity of all thought, including the most fundamental ideas of Marxism-Leninism.

This book is based primarily on newly accessible materials from Russian archives. It has also benefited from a rich set of books and articles on Soviet science and a growing body of work on postwar Stalinism.[25] Even before the opening of the archives, historians of Soviet science were in the vanguard of the study of late Stalinism. This can be explained in part by the desire to understand both the Soviet Union's tremendous scientific accomplishments—such as the rapid development of atomic weapons, the launching of the Sputnik satellites, and the steady stream of Nobel Prizes in science for work conducted during Stalin's time—as well as its equally noteworthy disasters, such as the outlawing of the study of genetics. Beginning in the 1960s and 1970s, a number of scholars, including Loren Graham, David Joravsky, and Alexander Vucinich used published materials and in some cases interviews and limited archival access to analyze the relationship between politics and science in the USSR. Their work furthered our understanding of scientific institutions, philosophical disputes in science, and the role of the state and Party in both supporting and suppressing scientific ideas.[26] Nonetheless, as David Joravsky noted in 1970 in *The Lysenko Affair*, a restricted source base forced him and his colleagues to "postpone the conventional first question of historical inquiry: Exactly which high-placed men got together with which others to effect this and that policy? That traditional method of beginning historical inquiry must await the opening of the archives."[27]

While the published materials clearly help frame the book, the narrative and analysis are based on precisely the materials to which Joravsky referred. In many ways, the subject was well suited for archival research because so many of the most important decisions were recorded by administrative sections and individuals whose papers are now declassified. For the postwar period the archives are well organized, reflecting a stability and efficiency within the Party and academic institutions that was missing in the 1920s and 1930s. Information about the discussions flowed up and down the bureaucracy, leaving a substantial paper trail that allowed me to piece together how decisions were made and how they were carried out. Papers in the Moscow Party Archive (TsAODM) and the Archive of the Academy of Sciences (ARAN) reflect the way disputes germinated and were dealt with among rank-and-file scholars. Papers in the Russian State Archive (GARF) and the Russian Economic Archive (RGAE) show how state organs took charge of implementing Party decisions and at times acted as the principal organizers of discussions. But the most valuable repository for understanding Stalin and science is the Central Party Archive (RGASPI), which contains the Central Committee papers including those of Agitprop, the Science Section, the Politburo, and the Orgburo and Secretariat. These documents, along with the personal papers of various Central Committee secretaries, record much of the organizational mechanisms for each of the discussions. As I was writing this book, more and more documents from Stalin's archive at RGASPI became accessible to researchers. These papers revealed in stark detail the extent to which Stalin became personally engaged with the scientific disputes.

Now that many of those archives are open to research, this book uses thousands of primary documents to show how the politics of science was practiced in the Kremlin by Stalin and his closest subordinates. It is not a traditional history of science in that the processes of scientific investigation, institutional development, and discipline formation are set aside so that politics and ideology can come to the fore. When background on the history of specific scientific fields is necessary, the book relies on existing disciplinary histories.[28]

This book branches out from previous approaches in three ways. First, it uses archival material to analyze six different discussions in detail, and thus avoids the temptation to extrapolate from one discipline to reach general conclusions about Stalinist science. This complicates our understanding of Stalin's motives for organizing debates and allows us to see how the approaches of scientists and the Party changed from one debate to another. No single discussion emerges as typical or paradigmatic. Second, the chapters pay careful attention to the ways in which shifting domestic political concerns affected decision making, arguments, and the grounds on which people defended their ideas. Scientific debates are understood as both a forum for political battles as well as a means of reaching ideological settlements that had effects far beyond the walls of academia. Finally, the book takes advantage of the recent declassification of Stalin's papers to place the "coryphaeus of science" at the center of the story. This material—which includes drafts of his essays, his extensive editorial comments on other people's written work, minutes from Kremlin meetings with scholars, and much more—reveals how these six discussions became focal points for Party politics and the effort to formulate Soviet ideology.

In many respects, Stalin's stint as the coryphaeus of science can be understood as part of the longer history of political leaders' desires to be taken seriously as thinkers. From Alexander the Great to the "enlightened despots" of the eighteenth century, heads of state have sought to justify their place atop the political landscape by placing their rule within a broader intellectual context. Confidence in the ability of human reason to control the natural and social environment blossomed throughout Europe in the century following the Enlightenment. Political leaders and political theorists alike held that the rational ordering of society based on the application of scientific knowledge would naturally lead to greater economic progress and social justice. By the twentieth century, governments in Europe and North America relied on rationality as a form of political legitimization.

Both superpowers in the Cold War claimed to have science on their side. In the United States, scientific administrators, such as Vannevar Bush and James Conant, and the sociologist of science Robert K. Merton argued that Western democracy and science mutually reinforced one another.[29] In 1950, Conant presented the mirror image of the Soviet argument: "Scholarly inquiry and the American tradition go hand in hand. Specifically, science and the assumptions behind our politics are compatible; in the Soviet Union by contrast, the tradition of science is diametrically opposed to the official philosophy of the realm."[30] Stalin also insisted on the unity of his political system and the scientific discoveries of his age. The effort to show how Marxism-Leninism constituted the best environment for science represents an extreme and at times brutal variation—but a variation nonetheless—on the broader story of the way in which science has been used to justify a full range of political systems in the modern world. In this sense, the story of the Soviet science wars offers lessons beyond the peculiarities of postwar Stalinism. Today's battles over stem cell research, global warming, and the teaching of evolution in schools are faint echoes of the controversies described in this book. To some extent, all modern societies must forge a working relationship between knowledge and power.

The physicist Peter Kapitsa wrote of the debate in his own field that "more than anything [it] reveals the mechanism of the Stalinist process. The battle of idealism and materialism in physics—this was only a philosophical mask which disguised political goals."[31] Like many of his fellow scientists, Kapitsa assumed that philosophy and politics were clearly distinct. But as the campaign for a coherent Marxist-Leninist ideology of science spread from one discipline to another, distinguishing the masks from the goals became difficult even for the participants, including Stalin. Philosophical content merged with political power; scientific argument melted into polemical leverage. The progress of science, which was so tightly intertwined with the self-image and foundational ideology of the Party, required procedures developed from within scientific and Party traditions. All the participants in the discussions appreciated the wider significance of their contributions. Their job entailed nothing less than the clear and forceful articulation of a worldview that placed the Soviet system at the pinnacle of historical development. Failure to accomplish this goal would undermine the Soviet Union's legitimacy for those living within its borders and for those observing the socialist experiment from around the world.

CHAPTER 2

"A MARXIST SHOULD NOT WRITE LIKE THAT"

The Crisis on the "Philosophical Front"

In late December 1946 Joseph Stalin called a meeting of high-level Communist Party personnel at his Kremlin office. The opening salvos of the Cold War had already been launched. Earlier in the year Winston Churchill had warned of an iron curtain dividing Europe. Disputes about the political future of Germany, the presence of Soviet troops in Iran, and proposals to control atomic weapons had all contributed to growing tensions between the United States and the USSR. Inside the Soviet Union the devastating effects of the Second World War were painfully obvious: cities remained bombed out and unreconstructed; famine laid waste to the countryside, with millions dying of starvation and many millions more malnourished.[1] All this makes one of the agenda items for the Kremlin meeting surprising: Stalin wanted to discuss the recent prizewinning book *History of Western European Philosophy*.

For Stalin, Soviet philosophy was both a scientific discipline and a political tool. Marxism-Leninism aspired to be more than a theory of politics and economics. It also claimed to provide a comprehensive worldview, known as dialectical materialism, which held that the material world provided the objective foundation of all knowledge.[2] Working on a research agenda set out by Marx and Engels and expanded by Lenin, Soviet philosophers were supposed to uncover fundamental and incontrovertible truths about the nature of human society and its evolution. They also served a political function, by engaging in debates with "Western," "bourgeois" philosophers who failed to understand the validity of Marxist doctrine and whose "reactionary" views might seduce unsuspecting people around the world. Stalin understood that there was a "philosophical front" to the Cold War.

Philosophy's double duty in science and politics was not new, of course. In the 1930s, the Party had equated certain philosophical positions with counter-revolutionary political activity. As a result, philosophers, like other social scientists, often shied away from what they thought were politically sensitive topics. Instead, most spent their time on tracts whose primary purpose was to provide academic justification for the Party's decisions and actions. The field became calcified. When Stalin turned his attention to the *History of Western European Philosophy* in 1946, he did so with the hope that philosophers could aid the party in addressing two striking features of postwar Soviet politics. First, since the mid-1930s, Soviet policy and rhetoric had become increasingly Russocentric.[3] Officially, the philosophical foundations of the system remained the same, but the Party also called on scholars to rewrite the history of philosophy to reflect the newfound role of Russia and Russians in the development of Marxism-Leninism. This goal went hand in hand with a parallel effort to downplay the significance of German philosophy in the Marxist tradition. These efforts were complicated by the fact that scholars could not admit openly that they were revising anything. Second, the battle with the West in the Cold War made philosophy all the more important. The Party had allowed work on Marxism-Leninism to slacken during the Second World War. Now Stalin called on philosophers to reinvigorate ideology for Soviet citizens and for use as a weapon in what he saw as a struggle with Marxism-Leninism's foreign detractors.

Georgii Aleksandrov, a participant in the December 1946 Kremlin meeting and the author of the philosophy book that Stalin had read and now wanted to discuss, appeared to be an ideal candidate for bringing together Soviet philosophy's scholarly and political strands. He had joined the Communist Party in 1928, when he was just twenty years old. As Soviet society underwent the wrenching transformations of forced collectivization and rapid industrialization, he studied philosophy at Moscow's Institute of Red Professors, a major training ground for Party cadres.[4] By 1933, he had graduated and begun teaching philosophy at another Moscow institute. As a young Party member in the 1930s he steadily gained responsibility, first on the executive committee of the Soviet-run Communist International and then, thanks to vacancies created by Stalin's purges, as the assistant to the head of the Central Committee's Department for Agitation and Propaganda. In 1940, he took over the department and with it responsibility for producing and spreading the Party's views on everything from international events to domestic activities in factories and on collective farms. Agitprop, as the department was known, also monitored scholarly activity, ensuring that teachers and researchers adhered to Party decrees. In February 1941, he became a candidate member of the Central Committee, and in March 1946 Stalin appointed him to the Orgburo, a small, elite Party committee chaired by Andrei Zhdanov, Aleksandrov's patron and the Party's second in command.[5]

Academic achievements had accompanied Aleksandrov's rise through the Party. In 1939 he earned a doctor of philosophy. In 1946 his *History of Western European Philosophy* received a Stalin Prize, the Soviet Union's highest scholarly award. And in November of the same year, when he was only thirty-eight years old, he was elected to the Academy of Sciences, the highest symbol of Soviet academic achievement.[6]

Despite his stunning success and the nearly universal praise his book had received, Aleksandrov had reason to feel apprehensive as he entered the red brick walls of the Kremlin and headed past the golden-domed churches toward the low yellow building that housed Stalin's office. His authority, so tangible when dealing with professors and his subordinates in the Central Committee, could easily evaporate in Stalin's presence. In the past year, he had been to the Kremlin six times to meet with Stalin.[7] He had witnessed enough to know that he too could be the subject of the leader's piercing criticisms. Though the press had lauded the *History of Western European Philosophy*, the importance of the subject meant that Stalin would pass the final judgment on the work's merits and faults.

Philosophy had caused problems for Aleksandrov in the past. During the war he just missed becoming a target of Stalin's ire when he and some of his senior colleagues wrote what they hoped would be a politically acceptable history of philosophy. The work's three volumes, published in 1940, 1941, and 1943 respectively, covered the history of European philosophy from antiquity to the nineteenth-century precursors of Marx.[8] Like Aleksandrov's 1946 book, this earlier three-volume work received a Stalin Prize—even before scholars and journals had the opportunity to review the third volume. It appeared that Aleksandrov had placed a feather in his cap. But by early 1944 other scholars had begun to attack the work, particularly volume III, which dealt with pre-Marxist German philosophy and utopian socialism. The authors had focused on those philosophers who, in their view, were precursors of Marx and Marxism. As one critic, Z. Ia. Beletskii, put it in a letter to Stalin in early 1944, the work's treatment of the German philosophers Kant, Fichte, and Hegel failed to identify their work as reactionary and bourgeois in nature. Since the Soviet Union was at war with German fascism, the letter continued, taking a conciliatory stand toward the German philosophical tradition meant losing a major battle along the philosophical front.[9]

Aleksandrov attempted to rebut the charges by explaining to his superiors in the Central Committee that Soviet scholars needed to reclaim the German roots of the Marxist tradition in order to challenge those philosophers who argued that German philosophy led directly to fascism.[10] He failed, however, to stem the tide of attacks on this history of philosophy, and in the spring of 1944 the Central Committee convened a series of meetings devoted to picking volume III apart. On May 1, the Politburo passed a resolution criticizing the volume as well as the field of philosophy more generally. The

Politburo declared that volume III glossed over the contradictions between the "progressive philosophy of Hegel's method" and his dogmatic philosophical system. Likewise, the book failed to illuminate the conservative nature of philosophy prior to Marx. Aleksandrov and his coauthors had erred in claiming to find the progressive, proto-Marxist elements in German philosophy and in challenging Western scholars' emphasis on the reactionary elements. Instead, they should have shown how Marx's ideas represented a radical break from German tradition.[11]

In the wake of the Politburo denunciation of the three-volume history, the Central Committee demoted Aleksandrov's coauthors, reorganized the Institute of Philosophy, and closed down the Soviet Union's leading philosophical journal. Central Committee secretary Aleksandr Shcherbakov told a group of propagandists that the Party was so concerned about the book's inadequacies because mistakes in philosophy often led directly to "political mistakes."[12]

From the late 1920s until the outbreak of the Second World War, Soviet scholarly disagreements had regularly turned into exchanges of bitter political denunciations with sometimes fatal consequences. The controversy over volume III gave an early indication that the postwar period would bring about a revival of the politicization of academia. But Aleksandrov, unlike his coauthors, managed to survive the Politburo's attacks unscathed, and he maintained his control over Agitprop. Still, in 1944, he admitted at a meeting of Agitprop employees that during the war their work had not gone well. Even as demands on his section increased, the conditions for improving propaganda and agitation remained strained by the war's devastation. *Pravda* was one-third its previous size, millions of cadres had died on the front, libraries had been destroyed during the German occupation, and there were fewer films and theaters to use for ideological purposes. Aleksandrov conceded that "facts are facts and we have reached the point where there is almost zero independent study of the work of Marxism-Leninism."[13] He saw his new book, the *History of Western European Philosophy*, as a scholarly means of reinvigorating work on the philosophical front.

Considering the battles over volume III, Aleksandrov must have understood that publishing his *History of Western European Philosophy* just three years later could put his career at risk. He knew the importance Stalin attached to philosophy and understood that his new work would be scrutinized at the very apex of Soviet power. Yet he had approached the topic with confidence, if not a fresh perspective: he wrote the book by piecing together notes he had made for lecture courses delivered in the 1930s. One of his colleagues suggested in a letter to the Central Committee that Aleksandrov now wrote about the controversial topic of German philosophy with such ease in part because he assumed that, with the end of the war, it was no longer necessary to distance Marx from Hegel and his other German antecedents.[14]

Being the head of Agitprop had its advantages. Reviewers praised the 1946 book, citing its militant tone, breadth of coverage, and attention to Party priorities.[15] Nonetheless, the debacle surrounding volume III of his previous work had given Aleksandrov ample warning of how quickly things could turn sour. In December 1946, as he made his way through the halls of the Kremlin office building, past the omnipresent guards, toward Stalin's office, he could only wonder what the leader's verdict would be. Was his new book worthy of the Stalin Prize it had received, or was it a failure suffering from the same politically inexcusable defects that had plagued volume III?

Aleksandrov was not the only person in attendance who was nervous about what would transpire at the meeting. In the spring of 1946 Peter Nikolaevich Pospelov—a powerful Party bureaucrat and chief editor of *Pravda*—had strongly endorsed the *History of Western European Philosophy* as the chair of the Stalin Prize Committee for science. If Stalin now decided that the book was somehow deficient, he too could lose the prestige and influence he had taken years to accumulate. Mark Borisovich Mitin faced a similar problem. Mitin had graduated from the Institute of Red Professors a few years before Aleksandrov and had also been one of the coauthors of volume III. Because he had been responsible for the sections on Hegel and German philosophy, criticism of the volume had affected him directly, and he had lost both his job as director of the Marx-Engels-Lenin Institute and his control over the philosophy journal *Under the Banner of Marxism*. These positions had made Mitin one of the most powerful academics in the country. Perhaps in an effort to rehabilitate himself, when the *History of Western European Philosophy* came out in 1946, Mitin nominated it for a Stalin Prize, writing that Aleksandrov's book was based on "profound scientific research" that revealed "a scientific understanding of the whole process of the development of philosophical ideas . . . [and] how prerequisites and conditions gave rise to the philosophy of Marxism."[16] What price would he pay if Stalin now determined that the book was no good?

Oddly, Zhdanov, Aleksandrov's boss and the Party secretary responsible for Agitprop, did not attend the meeting. The philosophy book certainly fell under his jurisdiction, but it is unlikely that Stalin snubbed him. The appointment of A. A. Kuznetsov to the Party Secretariat and the promotion of others from Zhdanov's clique show that his political fortunes were on the rise.[17] And Stalin had chosen Zhdanov to be the mouthpiece for the Party's campaign against ideological laxness. The previous summer he had led the charge against works of literature that Stalin believed had failed to live up to the Party's standards of engagement in ideological matters. His position as Stalin's chief assistant in the Party rested in no small part on his successful management of the ideological campaign. Surely, Zhdanov would have wanted to be among the first to know if Stalin had determined that yet another book on the history of philosophy had failed to meet his expectations. As it was, Zhdanov was not at the Kremlin to hear what Stalin had to say about Aleksandrov's book. His close associate, Kuznetsov, did, however, attend.

Figure 1. Minister of Education Sergei Kaftanov (*right*) awarding Georgii Aleksandrov (*left*) a Stalin Prize in 1946 for his book *History of Western European Philosophy*. Courtesy of Rossiiskii gosydarstvennyi arkhiv kinofotodokumentov.

Figure 2. A touched-up photograph of Stalin working at his desk in the Kremlin, November 1946. Courtesy of RGAKFD.

Stalin began the meeting by offering some instructions about other matters before turning to Aleksandrov's book, no doubt relishing the effect his words would have on those in attendance.[18] Then he launched into the book. He criticized it for being "divorced from the political battle" and for "lacking political spirit." He argued that Aleksandrov had failed to show the connection between Marxism and the people as well as the corresponding rifts between non-Marxist philosophers and the working classes. He complained about Aleksandrov's distant, academic tone, remarking that "a Marxist should not write like that." Ignoring Lenin's example of addressing philosophical controversies head-on, Aleksandrov had simply provided dry descriptions of various periods in the history of philosophy.[19] According to Stalin, the book did not explain the cultural and political context for the emergence of Greek philosophy or of the Renaissance.

Stalin claimed that the same problem was even more pronounced when it came to the book's description of German philosophy. It failed to show how Hegel contradicted himself, how his views exemplified "reactionary idealist philosophy," and that his main targets were the materialists of the French Revolution. Disagreement about Hegel's philosophy and the history of the dialectical method had dogged Soviet Marxists since the mid-1920s. When the Deborinites—who looked to Hegel as the focal point of dialectical philosophy—fell out of favor in the early 1930s, the younger philosophers who had taken their place were ill equipped to formulate philosophical ideas of their own.[20] As the denunciation of volume III and now Stalin's criticisms of Aleksandrov's book made clear, by the end of 1946 Soviet philosophers had still not integrated Hegel into the history of Marxist-Leninism in a way that satisfied the leader.[21] The book's failures, Stalin's comments emphasized, were both political and scientific. It neither adopted the right tone nor accurately gave an objective, Marxist interpretation of its subject.

If Stalin's goal had been to correct the book's mistakes or to diminish Aleksandrov's power, he would have had plenty of direct means at his disposal. He could easily have instructed Zhdanov or Kuznetsov to draft a decree denouncing Aleksandrov, removing him from his post at Agitprop, and nullifying the book's Stalin Prize. He could have signed a Politburo decree criticizing the book, or commissioned an article to do the same thing in *Pravda*. But despite his clear distaste for the book, Stalin did not make his views known to the public or restructure Agitprop. Instead, he ended the meeting, which had lasted over an hour, by ordering the Institute of Philosophy to organize an "open discussion" of the book.[22]

As was often the case during Stalin's reign, the leader's subordinates scrambled to turn his ideas into actions. On December 26, 1946, the Central Committee Secretariat passed a resolution on the book, noting that it contained serious errors and calling for a meeting in early January at the Institute of Philosophy that would "ensure complete freedom of criticism and exchange of opinions about the book." The new Party secretary Kuznetsov took responsibility for overseeing the meeting, with one of Aleksandrov's Agitprop deputies handling the organizational details. Reflecting the general interest in the meeting and its importance to the Party and the entire academic community, the resolution ordered that three different journals plan to publish the proceedings and that prominent members of the Party, press, and academic organizations join the audience. As the date of the meeting approached, Kuznetsov increased the number of Central Committee bureaucrats attending and gave the responsibility of opening the discussion to V. S. Kruzhkov, the director of the Marx-Engels-Lenin Institute, who had attended the meeting at Stalin's office. He also summoned a group of philosophers to the Central Committee, suggesting that they participate. Meanwhile, Stalin's personal assistant, A. N. Poskrebyshev, kept abreast of Kuznetsov's plans and kept his boss up to date on what was happening.[23]

The Institute of Philosophy's discussion of Aleksandrov's book began on the evening of January 14, 1947. Nearly 400 people attended the session, including 68 from the Central Committee, 40 from the press, 57 from the presidium and institutes of the Academy of Sciences, and 137 from the Central Committee's Academy of Social Sciences, Moscow State University, the Marx-Engels-Lenin Institute, and other scholarly institutions.[24]

Stalin's criticisms and the Central Committee resolution sponsoring the discussion notwithstanding, the organizational structure of the January meeting favored Aleksandrov. The stated purpose of the meeting was to criticize his book, but the participants in the discussion varied in their interpretations of the meaning of Stalin's remarks and the extent to which the book and its author should be held responsible. The number of people who had attended the meeting in Stalin's office, and therefore had firsthand knowledge of what had been said there, was limited. Besides Aleksandrov, only five other people who had been at the Kremlin meeting in Stalin's office planned to speak at the January discussion: Kruzhkov, F. D. Fedoseev, M. T. Iovchuk, Pospelov, and Mitin. Kruzhkov, in his capacity as the director of the Marx-Engels-Lenin Institute, reported to Aleksandrov. Fedoseev and Iovchuk were Aleksandrov's deputies in Agitprop. Pospelov, the editor of *Pravda*, was not directly responsible for philosophy and had little incentive to upset the status quo. And Mitin had to deal with the fact that he had endorsed Aleksandrov's book for a Stalin Prize. Most importantly, Aleksandrov clearly outranked each of them. As long as he remained head of Agitprop, even those who had heard Stalin's remarks had to be very circumspect in voicing criticisms of their own.

The three-man presidium that chaired the meeting posed even less of a threat to Aleksandrov. Kruzhkov, Vasetskii (director of the Institute of Philosophy), and Academy of Sciences president Vavilov had little to gain from ridiculing Aleksandrov. (Vavilov, who had spoken in defense of the book at the Stalin Prize deliberations, sidestepped responsibility at the meeting by failing to attend after the first day.)[25] In the whole country, only Kuznetsov, Zhdanov, and Stalin were Aleksandrov's superiors when it came to questions of ideology. Unless one of these three delivered the concluding speech—and this did not seem to be in the cards—Aleksandrov was the most powerful political figure participating in a conference organized to criticize his scholarship. Less prominent philosophers were in no position to confront Aleksandrov. As one of them put it, Aleksandrov "had the fortunate opportunity which many of us do not have—to receive direct instructions from comrade Stalin."[26] As long as he had a special access to Stalin, Aleksandrov seemed secure against challenges from below. Of course, a few other speakers could refer to the meeting with Stalin, but there was no consensus on what the "complete freedom of opinions and exchange of criticism" was supposed to accomplish. Was the discussion solely about Aleksandrov and his book, or was the whole field of philosophy open to criticism? Was Aleksandrov a target only as a scholar, or could his stewardship of Agitprop also be attacked?

Opening the discussion, Kruzhkov clarified that Stalin had pointed out "shortcomings and mistakes" in Aleksandrov's book. He emphasized the book's lack of "militant Marxist spirit" and its "passionless academic judgment." Still, he assured the audience that it was an improvement on any previous history of philosophy, including the ill-fated volume III. As he saw it, the problems ranged beyond Aleksandrov, to include the reviewers of the book and the efforts of philosophers in general, who had not responded to Zhdanov's call to activism among the intelligentsia.[27] This established a crucial line of defense for Aleksandrov: since Stalin had not mentioned any problems with Aleksandrov's control of Agitprop, the more that the discussion concerned the field in general, the less likely he was to face unconstrained criticism. Furthermore, those who had been discredited by volume III, such as Mitin, had to tread lightly for fear that their own errors—which apparently had been more serious—would once again be mentioned.

True to form, when Aleksandrov addressed the meeting, he emphasized that philosophers in general had not responded with enough vigilance to the Central Committee rulings about volume III or Zhdanov's speeches on ideology. As far as the errors in his own book were concerned, he noted that he had spoken with Stalin and agreed that he would correct them. The two of them, he implied, had already decided what revisions were necessary—his fate and the fate of the book would be decided by Stalin himself, not by this, or any other, discussion. To his mind, the purpose of the meeting, then, was to strengthen the field of philosophy more generally and to bring it in line with the broader ideological campaign.[28] Pospelov spoke third, and though he mentioned the book's shortcomings—it lacked Party spirit, it was too objective, it failed to show how Marxism's scientific foundations distinguished Marxism from all previous philosophical schools—he also praised the book's strengths and claimed that with some revisions it could serve as the basis for a textbook for students.[29]

The keynote speaker, the head of Agitprop, and the editor of *Pravda* had all offered a united, and moderate, interpretation of Stalin's criticisms. After them, few of the rank-and-file philosophers dared to mount substantial attacks. There were a few exceptions. Mitin was one. He had attended the

December meeting with Stalin at the Kremlin and believed that Stalin's statements had been unambiguous. Mitin argued that Aleksandrov had failed to learn from earlier criticism and therefore his book was as bad as, if not worse than, volume III. It represented a "complete disaster" on the philosophical front.[30]

Z. Ia. Beletskii, a professor of philosophy at Moscow State University, took things even further, suggesting that Aleksandrov's views placed him squarely in the "Menshevik-idealist" camp. By reviving an epithet from the 1930s, Beletskii attempted to disrupt the collegial tone of the meeting. It was not a matter of fixing this or that part of the book; the whole thing was rotten. Beletskii's fearlessness in attacking his Party superior may have stemmed from the fact that his letter to Stalin in 1944 had precipitated the controversy surrounding volume III and a similar letter to the leader in 1946 highlighted problems with Aleksandrov's book. Though he had not attended the Kremlin meeting, he thought that the point of the present discussion was to address the obvious problems with Aleksandrov's book and not to offer mild criticisms of the field of philosophy in general.[31]

Subsequent speakers, however, rebutted Mitin and Beletskii's points. Aleksandrov's mistakes, they argued, were all too common in Soviet philosophy, and therefore everyone had to take some responsibility.[32] B. M. Kedrov, a philosopher of science and assistant director of the Institute of Philosophy, led this counterattack. Rejecting Mitin's assertion that the book was a "complete disaster," Kedrov reminded the audience that "volume III was a harmful book" that could not be redeemed, whereas Aleksandrov's book remained the basis of further work. He added that Mitin had failed to undergo the proper degree of "self-criticism" in regard to his own mistakes. Self-criticism was a mainstay of Bolshevik rituals, in which speakers publicly repented for views that had been subsequently denounced by the Party. As Kedrov put it to the amused audience, Mitin understood self-criticism as "wait until your mistakes are repeated by someone else and then bravely, morally, and without fear criticize your own mistakes . . . without mentioning your own name in the process."[33] The vast majority of the subsequent speakers sided with Kedrov and dismissed Mitin's comments in part because he had failed to openly discuss his own responsibility for the failures of volume III.

Fedoseev, Aleksandrov's assistant at Agitprop, took on Beletskii, noting that the attendees had not gathered to "crush or even humiliate someone" or to "gloat and giggle because a book came out with deficiencies and mistakes." He ridiculed Beletskii's demagogic tone and his claim that Aleksandrov had written from an anti-Marxist point of view.[34] As it turned out, neither Mitin nor Beletskii found many colleagues who were willing to support them at the meeting.

Clearly, what was said, when, and by whom influenced the outcome of the discussion. This was not simply a matter of ritually arriving at foregone conclusions. The minutes of the meeting reveal that personal politics, rhetorical style, and organizational structure all altered the outcome. Even though Stalin had expressed his views of the book, the direction of the discussion at the Institute of Philosophy remained flexible. Scholars did not intentionally undermine Stalin's wishes so much as they interpreted and tried to shape his rulings in ways that were favorable to the institutional, political, and even intellectual positions they held.

The subjects ranged far beyond those Stalin had brought up at the Kremlin. Some participants drew on the rising Russian nationalism of the 1940s to question the book's emphasis on Western Europe at the expense of Russia. In their view, Russian thought played a central role in the history of philosophy that Aleksandrov had patently failed to explore.[35] Others complained about the field in general, noting the overall lack of productivity, especially among those who did not hesitate to criticize the work of others. One participant discerned two different approaches to scholarship in the USSR. The work that was published was often weak, "monotonously dry and superficial" while another "richer, varied and deeper" set of monographs, dissertations, and articles never made it into print. The obstacles to publications seemed to favor simplistic and popular brochures and articles. Only the true "democratization" of the field could wrest control from the leaders and allow truly scientific work to see the light of day.[36] Others argued that a new journal dedicated specifically to philosophy would help save the field from popularizers and return it to the scholars.[37]

Aleksandrov's closing statement was for the most part measured and confident—even patronizing. He claimed that he was pleased with the way the group had taken the Central Committee's concerns seriously and was glad that the meeting had produced what he believed to be a thorough discussion of philosophy. He also singled out Beletskii, noting that "Comrade Beletskii has written nothing in his life and does not want to write anything and does not want to defend the Marxist tradition." He had not earned the right to criticize other people's work.[38]

Kruzhkov closed the discussion by declaring it a success. Though he suggested it was not a time for "polemics," he also registered his displeasure with Beletskii, whose views could "in no way be considered Marxist." He emphasized that the Central Committee would judge the work of all philosophers based on their development of "powerful ideological tools in the struggle against everything hostile to our Soviet ideological strength." Declaring his confidence that future work on the philosophical front would be successful, Kruzhkov pronounced the discussion of the book closed.[39] Aleksandrov had apparently survived without serious damage to his reputation as a Party leader or a scholar. If he was concerned by the fact that, beyond general statements, nothing had been done to aid the effort to write an acceptable textbook on the history of philosophy, he did not show it.

Kruzhkov's initial reports to Party secretary Kuznetsov on the meeting at the Institute of Philosophy repeated that it had been a success. The participants had criticized Aleksandrov's book, while revealing shortcomings in the work of the Institute of Philosophy, in the Philosophy Department of Moscow State University, and in publications in the field. Kruzhkov also noted Mitin's lack of self-criticism and Beletskii's "incorrect and demagogic tendencies." In conclusion, he wrote, the proceedings of the discussion should be published as "an instructive example [showing] how our Party, and comrade Stalin personally, teach how to conduct the brave, consistent principle of applying Marxist-Leninist party spirit [*partiinost'*] in all our ideological work." He also sent Kuznetsov a draft of a report addressed to Stalin about what had transpired.[40]

Stalin did not rely solely on Kruzhkov's report to learn about the discussion. One participant in the conference picked out someone in the audience whom he did not recognize and who did not seem to be pleased with what was going on. As the observer later recalled: "In the first row of the large hall sat a man in a general's uniform, expressing indignation from time to time in the form of gestures and retorts about different speeches and also about the nature of the chairman's handling of the discussion. This man, they told me, was Stalin's secretary Poskrebyshev. Probably, he reported to the 'boss' about the 'liberal' character of the criticism."[41] Kuznetsov's response mirrored Poskrebyshev's. He wrote on top of Kruzhkov's positive report, "There is little objectivity" and crossed out a number of sections, including any negative remarks about Mitin and Beletskii, Aleksandrov's two most vocal critics. Eventually the draft was simply put aside. Meanwhile, on February 7 Kruzhkov sent Kuznetsov another review of the discussion, this time in the form of a draft of what would be published in the press. Kuznetsov forwarded it to Zhdanov and others in the Secretariat, but his notes indicate that he was still not pleased with the results. Among other things, the draft failed to mention that Stalin had initiated the discussion, and it spent too much time praising the book's merits. So Kruzhkov submitted another draft. It too was rejected. Finally, on March 14, the Central Committee looked into the results of the discussion and passed a resolution requiring Kruzhkov to write yet another draft for publication in two days "taking into account the exchange of views" that had been voiced at the Secretariat. On March 22, this draft went straight to Zhdanov, who had now taken direct control of the situation.[42]

The new report reflected Stalin's initial criticism of the book, noting his contention that it contained specific interpretive mistakes about German philosophy, that it lacked political militancy, and that it was important to understand the discussion as part of the effort to invigorate work on the ideological front.[43] Despite this newfound rigor, Kruzhkov's final report was never published. Instead, the Politburo took over the issue from the Secretariat. Dis-

Figure 3. Members of the Politburo cross the territory of the Kremlin on their way to the May 1 celebration in 1947. *Left to right*: Nikolai Voznesenski, Georgii Malenkov, Andrei Zhdanov, Lavrenty Beria, and Viacheslav Molotov. Courtesy of RGAKFD.

pleased with the results, Stalin had decided to take another approach to correcting Aleksandrov's mistakes.

On April 22 Stalin and the Politburo passed a decree titled "On the discussion of c[omrade] Aleksandrov's book *History of Western European Philosophy*." After reviewing the materials from the January meeting at the Institute of Philosophy, Stalin had decided that the "organization of the discussion as well as the method of handling its results turned out to be unsatisfactory," which made the whole endeavor "limited [and] ineffective." The Politburo resolved to organize a new discussion in the late spring, with participants from all over the Soviet Union and with all the speeches recorded by a stenographer and published. With Zhdanov now in charge of the details, the discussion became a nationwide affair and a strategic piece of the broader *zhdanovshchina*.[44]

The "guest list" of the second discussion of Aleksandrov's book reads like a "who's who" of late Stalinist politics and scholarship. Twenty-three members of Agitprop attended, including the heads of nearly every section within it, as did dozens of Central Committee bureaucrats and the assistants to Politburo members Georgii Malenkov and Lavrenty Beria. Stalin's assistant Poskrebyshev was there, no doubt preparing a report of his own to be submitted directly to Stalin. The Moscow Party organization, the Red Army, the Union of Soviet Writers, the State Publishing Administration, and major presses sent representatives. Professors and academics in fields ranging far beyond philosophy also appeared on the guest list. On average over three hundred members of the Party and the scholarly elite attended each of the eight sessions between June 16 and June 25, 1947. The venue for the meeting also changed: instead of being held at the Institute of Philosophy, the second discussion took place at the Central Committee building on Staraia Ploshchad'.[45]

This time the organizational structure did not favor Aleksandrov. Zhdanov presided and prevented Aleksandrov and his allies from dominating the opening statements. Still, Zhdanov was coy about the direction he thought the discussion should go. He opened the meeting with only a brief announcement noting, without going into specifics, that the Central Committee had not been pleased with the first discussion. He did not mention Stalin or his specific criticisms. Instead, he invited the speakers to discuss Aleksandrov's book honestly. To emphasize the open and egalitarian nature of the meeting—or at least the appearance of such openness—he refused to allow a prominent philosopher or Party leader to give an agenda-setting talk, turning the lectern over instead to a relatively unknown scholar from Leningrad.[46]

Without greater direction from Zhdanov, participants were uncertain about the range of topics the leadership expected them to discuss. After the first speech, Zhdanov received a note from the audience asking him whether the speakers should restrict their comments to Aleksandrov's book or expand them to consider questions related to the philosophical front in general. Zhdanov remained ambiguous: "If the comrades mean to touch on questions about the situation on the philosophical front which are not to the detriment of the main theme, we will not restrict them." An inquiry into how long the discussion would last received a similarly open-ended response: "The Central Committee is prepared to give as much time to this discussion as you need."[47]

As it turned out, Zhdanov delivered his speech—the event that would come to define the meeting and serve as a major reference point for the whole postwar ideological campaign—on the ninth day and seventh session, after forty-three others had already addressed the assembly. Rather than enforcing a Party line in philosophy, the discussion helped to forge one: Zhdanov's speech responded to ideas put forth by other, lower-ranking participants. Certainly, anticipation of the discussion's outcome influenced what many speakers chose to say, while others simply used the forum to air old disagreements. But the majority undertook efforts to interpret Stalin's wishes and to improve the situation in philosophy in general and with Aleksandrov's book specifically. Almost all of the discussion took place without a clear statement outlining what Zhdanov or Stalin expected.

Confusion and disagreements concerning the history of Marxist thought suggested that the contours of Soviet ideology were far from clear. But participants understood that the rewards for settling these issues would be high: a truly Soviet Marxist interpretation of the history of philosophy, they believed, would be a weapon in the expanding effort to win the support of people around the globe. Grasping that the Cold War struggle was spreading to Eastern Europe and beyond, philosophers called on each other to show the supremacy and veracity of the Soviet worldview. The Second World War had ended with the defeat of fascism; now the emphasis shifted to winning the peace.

These lofty goals help explain the Party's intense scrutiny of incompetence and lack of productivity among philosophers. The situation required more than another speech by Zhdanov. Instead, he played the role of judge, offering verdicts about the issues that others raised in the course of the discussion. As the others spoke, he made notes that he then used to draft his speech. On June 23, he sent a draft to Stalin with a note "urgently begging" for the leader's "instructions." He planned to deliver the speech to the meeting on the next day. Then he would let the discussion continue for another day or so before concluding it. Stalin responded that the speech "had turned out fine," but suggested some structural changes and editorial corrections.[48] The exchange makes it clear that Zhdanov wrote the speech and Stalin approved it in the midst of the discussion, not in advance of it.

Three broad and interconnected themes developed during the course of the discussion and Zhdanov's response. The first concerned the relationship of science to politics. To what extent was Soviet philosophy—or any other form of scholarship—responsible to the immediate demands of the Party and to what extent was it "objective" and factual and therefore beyond political concerns? Attempts to answer this question quickly led to disputes about the audience for philosophers' articles. Were philosophers supposed to write primarily for one another and work on solving academic problems, or were they supposed to write for a general public in a manner that would combat "bourgeois" ideas on the "philosophical front"? A second, related theme had to do with the proper sources for philosophical truth. Did the canonical texts of Marx, Engels, Lenin, and Stalin provide the guidelines for all legitimate philosophical inquiries? Or was philosophy supposed to evolve through the careful observation of the objective world much as the natural sciences did? Finally, a third theme concerned the role of Russia and Russians in the history of Marxism, which logically resulted from the decision to downplay the influence of Hegel and other German philosophers on Marx. How could Marx be integrated into the expanding Soviet propaganda about the primacy of Russia in world history? These three themes surfaced in response to Aleksandrov's book and the situation on the philosophical front. Because the philosophy discussion failed to give concrete answers to these questions, the themes also reemerged in other contexts in subsequent debates about ideology and knowledge in the USSR.

Mitin, who must have felt vindicated by the failure of the January discussion, led the charge against Aleksandrov's lack of Party spirit and of militancy. He reiterated that the purpose of Soviet philosophy was to engage opposing opinions in the spirit of battle, not to hide behind the veneer of academic objectivity.[49] Others chastised Aleksandrov for shying away from "political engagement" and for failing to "combat bourgeois histories of philosophy." Writing philosophy was fundamentally a political act, they argued. A textbook needed to be a weapon for the Party and could not simply be written for professional philosophers.[50]

The problem was that nobody could identify a work of philosophy—other than the writings of Marx, Engels, Lenin, and Stalin—that had been sufficiently rigorous and politically engaged. But if philosophy was fundamentally about politics, who—besides luminaries combining political and scholarly authority—could be trusted to write an acceptable book? To many the answer seemed to be Aleksandrov, who, after all, was the head of Agitprop and a true political insider. Despite his book's problems, many of the participants conceded that it was "a step forward" and it was "possible to recommend [it] as a popular textbook."[51]

In his speech, Zhdanov echoed Mitin's concerns about Aleksandrov's lack of political engagement, denouncing the "cold, indifferent, and objective" use of facts. He compared Aleksandrov to a "preacher of toothless vegetarianism" who managed to "say nice things" about all philosophers. In Zhdanov's opinion, this lack of "militant spirit" characterized Soviet philosophical work in general and not just Aleksandrov's book. He noted, "The phrase 'philosophical front' has often been used here . . . But where, strictly speaking, is this front?" Instead of evoking a "detachment of military philosophers, fighting for the perfection of Marxist theory, leading the decisive blow against hostile ideologies abroad," Zhdanov saw a "quiet factory or an encampment somewhere far away from the field of battle."[52]

Why had Aleksandrov not used a more militant tone in writing his *History of Western European Philosophy*? A quick review of articles he penned in the Soviet press from 1945 to 1947 shows that he had no trouble emphasizing Party goals and using the language and polemics that Zhdanov was demanding. The answer comes from the assumption, shared by Aleksandrov and many others, that there was a distinction between propaganda and serious, academic work. Clearly, he thought an "objective" style seemed appropriate for a textbook. V. I. Svetlov, the assistant minister of education and a former director of the Institute of Philosophy, sympathized with Aleksandrov's choice. He warned that philosophers were becoming "publicists" who wrote for popular journals but did not advance their field. Increasingly, they were drawn into work in the Party and state apparatus, which distracted them from their scholarly responsibilities.[53] As another speaker argued, "political passions" that were key to Party work were "out of place" in a textbook.[54]

Others came to the defense of the "publicists" and argued in favor of an engaged style for philosophy in general and for its textbooks. An editor of *Pravda* noted that shying away from "publicity work" had been Aleksandrov's biggest mistake. Work in the press, propaganda, and Soviet philosophy were all part of a single effort. There was no point in dispatching a group of philosophers to do battle with "reactionary bourgeois philosophy" while everyone else continued their scholastic arguments in the rear.

Rather than adapting their styles to meet the demands of popular publications, many speakers expressed their desire for a journal dedicated solely to philosophical issues. When *Under the Banner of Marxism* ceased publication in 1944, philosophers no longer had their own journal. In 1946, some philosophers struggled to rectify the problem. They had in mind a publication that could review and discuss ideas that needed a wider audience than could be found at the institutes and universities but less attention than the Central Committee–approved and politically saturated articles that appeared in *Bolshevik* received. In 1946, the president of the Academy of Sciences and director of the Institute of Philosophy sent a letter to Zhdanov and Malenkov asking for a journal specifically designed for publishing philosophical scholarship. Zhdanov sent the letter on to Aleksandrov, who supported the idea of a philosophy journal, adding that it would assist in the effort to counteract the "considerable work on questions of philosophy" being produced at the time in Western Europe and the United States. He drafted a Central Committee decree establishing a philosophy journal, but the project did not make it off the drawing board.[55]

Bolshevik, the nationwide publication that most often addressed philosophical issues, was the theoretical and political organ of the Central Committee, which left little room for airing discipline-specific arguments, let alone debate. Philosophers felt that at times their ideas needed to be worked out among themselves before going into *Pravda* or *Bolshevik*. Without an alternative to these outlets, some scholars refrained from submitting work for publication. The Central Committee's decision not to create a philosophical journal in 1946 may have indicated its apprehension about allowing philosophers to spend too much time addressing questions that might not be central to Party politics and ideology. Leaving *Bolshevik* as the main organ for philosophy meant that the Central Committee had direct control over subjects and arguments in the field. While Party officials argued that not enough work was being produced in philosophy to justify a philosophical journal, philosophers countered by claiming that the absence of an appropriate organ for their work had diminished the output of high-quality philosophy.

At the January 1947 discussion of Aleksandrov's book, the subject of a philosophy journal had resurfaced. Kedrov, an ally of Aleksandrov and an opponent of Mitin, had been one of the proposed journal's main supporters. Whereas Mitin had called for more militancy and Party spirit, Kedrov had sought a forum where such a tone might be less necessary. Another participant had argued at the time that "like air, we need the widest discussion of philosophical questions. ... we need ... a Marxist-Leninist philosophical journal."[56] Following the January discussion, the Academy of Sciences president again wrote to the Central Committee asking permission to create a philosophy journal. But with the plan for the larger Central Committee discussion in the works, the Secretariat did not act on the project.[57]

At the June discussion Kedrov appealed directly to Zhdanov, explaining that a journal would invigorate work in the field, educate young cadres, and help to avoid repeating the troubles that now surrounded Aleksandrov's book.[58] Svetlov took up the argument as well, noting that *Bolshevik* had turned down articles because they had not been politically vigorous enough for the journal. Scholars responded by not submitting their work. At this point Zhdanov interrupted Svetlov by suggesting that good articles would get published in existing journals. But Svetlov rebutted by noting that "not all articles are accepted, and not because they are bad, but because . . . [some people held the opinion that they] are impossible to put in *Bolshevik* because they will not be interesting for a wide readership and will not find an audience." Zhdanov could not understand why, if philosophers were really being forced into popular work, there were still so few articles in *Bolshevik*. Svetlov could only reply that it was because "publicist-philosophers are in short supply."[59] Zhdanov wanted to define philosophy in terms of its contributions to the Party and to educating the masses; those in favor of an independent philosophy journal saw their field as an academic discipline, despite its politically volatile subject matter.

By the time Zhadnov took the floor, he had checked with Stalin about the journal. With the leader's approval, he noted that "the current opportunity to publish original monographs and articles is not fully exploited." He also directly confronted Svetlov's claim that the specialized work of philosophers would not be appropriate for *Bolshevik*, an opinion that he attributed to "an obvious underestimation of the high level of our audience and its aspirations." Philosophy, he continued, was not "entirely the property of a small circle of professional philosophers, but . . . in fact the property of all our Soviet intelligentsia."[60] Zhdanov's comments made clear that there was no room for a journal dedicated exclusively to philosophical work. At least in print, philosophy was entirely subsumed under ideology. And without their own journal, it appeared at this point that the philosophers needed to go the route of "popularizers" and "publicists" to earn the Party's full support. Philosophy was being made a subset of Party propaganda.

One reason philosophers were so persistent in their demand for a journal was that they believed that both lingering and newly emerging philosophical problems could not be worked out in the popular press. A number of these problems centered on unresolved disputes about the relationship between philosophy and science. Because the Party's position on controversial questions in the philosophy of science had not been established, no amount of militancy could solve the problem. The Party itself needed the help of philosophers and scientists in order to clarify the issues. Scholars' first impulse was to turn to the works of Marx, Engels, Lenin, and Stalin. Engels's *Anti-Duhring* and *Dialectics of Nature* and Lenin's *Materialism and Empiriocriticism* provided some clues about the integration of science and Soviet philosophy. But few philosophers had the technical background to articulate a tenable philosophy of science that could encompass both dialectical materialism and, for instance, quantum mechanics or modern genetics. A rift seemed to develop between the natural scientists and the philosophers. The fact that philosophers and scientists in the West were also exploring precisely the relationship between science and philosophy compounded the problem. If Soviet philosophers could not put forth a Marxist-Leninist theory of science, the scientific intelligentsia would be susceptible to the "idealist" and "bourgeois" theories of their Western counterparts.

There seemed to be two possible solutions: natural scientists could be encouraged to take philosophical issues more seriously, or philosophers could learn more about the natural sciences. On the first day of the discussion, Kedrov, the head of the Philosophy of Science Section at the Institute of Philosophy, complained that Aleksandrov had not given enough attention to the development of the natural sciences and their vindication of the materialist worldview.[61] Another speaker complained that because Aleksandrov's analysis ended with the dawn of Marxism, he had not explored the important developments in the natural sciences that had occurred since Engels wrote *Dialectics of Nature*. New developments in atomic physics posed particularly vexing problems, as they appeared to challenge the materialist principles at the heart of Marxist-Leninist theory.[62]

A number of speakers singled out Aleksandrov's inadequate explication of the history of the natural sciences, calling for general improvement among philosophers in their knowledge of the natural sciences. Philosophers, apparently, received little to no advanced training in mathematics and physics. Historians of philosophy also needed to develop the tools to understand how discoveries in the natural sciences related to philosophical advances.[63] One speaker plainly noted that if "philosophical workers are called to synthesize a development in the field of natural science, then they should be no less well prepared than the specialists in the fields of theoretical physics and chemistry." But Zhdanov interrupted to suggest that it was the scientists themselves who were deficient: "Are you sure that our specialists in chemistry are well grounded in the field of scientific-materialist philosophy?"[64]

According to the accepted Soviet theory of science, the natural sciences were not supposed to progress independently of a materialist worldview. One philosopher of physics, from Kiev, argued that science did not develop according to some sort of internal "logic" based on "thought." Instead, he held that economic and political forces contributed to the progress of both philosophy and science. "Bourgeois," "idealist," foreign philosophers, he emphasized, were using analyses of modern science as weapons on the philosophical front.[65] Another philosopher of physics addressed the contemporary problem of combating the idealist interpretations of the Copenhagen school of physics, which he feared were infiltrating Soviet textbooks. Young physicists who wanted to learn modern physics from a dialectical materialist point of view were unable to do so.[66]

In his speech Zhdanov asserted that all scientific knowledge was subject to the Party's interpretation. He argued that Aleksandrov had erroneously "separated [the history of philosophy] from the history of natural science" and failed to recognize the materialist foundations of scientific progress. The problem went much deeper than Aleksandrov's book. "Modern bourgeois science," Zhdanov continued, propagated all sorts of mistaken philosophies. He chastised the followers of Einstein for their confusion of absolute and relative truth, noting: "The Kantian vagaries of modern bourgeois atomic physicists lead them to inferences about the electron's possessing 'free will,' the replacement of matter with some sort of collection of energy-bearing waves, and to other devilish tricks."[67]

Zhdanov also asserted that philosophers had the enormous responsibility of integrating and analyzing the findings of modern science, keeping in mind Engels's declaration that materialism had to progress with each great discovery in the natural sciences.[68] So the hard sciences were to be both sources for philosophical growth and by-products of dialectical materialism. Zhdanov remained concerned that some Soviet scientists were idealists, influenced by bourgeois conceptions. Philosophers could help scientists by battling against bourgeois philosophies of science abroad and combating their influence within the borders of the Soviet Union. Science could help philosophy as well: scientific advancements aided the progress of dialectical materialism and Party ideology. Science, like the works of Marx, Engels, Lenin, and Stalin, provided the "objective," material basis for philosophical work.

Since philosophy itself was a science, it too had to be protected from negative influences. In Aleksandrov's case, the problems originated with his attempt to come to terms with Hegel's influence on Marx. One group of speakers at the June discussion simply dismissed Hegel's influence as reactionary. This, in turn, caused others to come to his defense, noting that Stalin's depiction of Hegel as "aristocratic and reactionary" did not mean that parts of his work were not crucial to Marx's development. Still others claimed that Hegel's role in Marxism was essential. One philosopher from Leningrad asked, "If Hegel's philosophy was completely reactionary," how could "such smart and insightful geniuses such as Marx and Engels [have] slipped into a clever, open trap?"[69] But Zhdanov characterized the whole discussion of Hegel as "scholastic" and "unproductive." Philosophers should be addressing contemporary philosophical problems, not the German origins of Marxist thought.[70]

If German philosophy was taboo, what was the proper subject for the history of philosophy? It was easiest to identify what did not belong. One speaker from Estonia suggested that all Western European philosophy was suspect. He noted that the bourgeois worldview had been taught in Estonia for years and that, during the three years of German occupation, fascists propagated their theories of racism. He ridiculed a strong and troubling tendency to look to the West for guidance, kowtow to Western ideas, and even to teach Fichte, Kant, Dewey, James, Spencer, Nietzsche, and other non-Marxist philosophers. Aleksandrov's book, he complained, provided no help in counteracting this trend.[71]

As a number of speakers pointed out, the category of "Western European philosophy" constituted a major problem with Aleksandrov's book. A philosopher from Tashkent lamented the lack of attention to the East and the undifferentiated understanding of "Arab philosophy" when it was mentioned at all. Another from Erevan accused Aleksandrov of adopting bourgeois tradition by ignoring Byzantine thought. Teachers wanted a history of philosophy that would help their students struggle with bourgeois historiography. One scholar, displaying a literal understanding of Aleksandrov's title, thought that the book should really have started with the Romans since "Greece is in southeastern [and not western] Europe." Another speaker thought that the title revealed a political mistake, since it accepted the false and arbitrary division between East and West.[72]

But the safest way to approach the history of philosophy was to elevate the role of Russians. Playing to the patriotic fervor of the period, a number of philosophers argued that the Russian philosophical tradition needed to be included in any Soviet history of European philosophy. While one philosopher lamented the exclusion of the "brotherhood of Slavic peoples" in general, most who spoke to the issue were emphatic about the need to include Russians specifically. As the argument went, instead of "kowtowing to the West," Aleksandrov should have stressed the international significance of Russian thinkers, including Lomonosov, Radishchev, the Decembrists, Belinsky, Herzen, Chernyshevsky, and the Russian Marxists. This was emblematic of the lack of "Russian patriotism" among Soviet philosophers on the whole.[73]

Another speaker pointed out that work on the history of Russian philosophy and propaganda about Russian philosophy had recently received "colossal support," and "masses of our people heard for the first time about our great Russian thinkers." But the job of incorporating this propaganda into scholarly books remained unfinished.[74] Kedrov tried to temper this nationalist wave by reminding his listeners that national categories were not as decisive for historical and philosophical development as class categories.[75] Svetlov advanced a compromise position that accepted some degree of Russocentrism but not at the detriment of class: Soviet people should be proud of Belinskii, Herzen, Chernyshevsky, and others, but they should know that these thinkers were not "scientific philosophers" like Marx. Throughout the discussion, there was a general push to argue that these Russian thinkers, because they were progressive, deserved as much attention in a history of philosophy as their nineteenth-century German counterparts. Russians, not Hegel, proved decisive in shaping Marxism.

While many spoke of integrating the history of Russian thought into the history of Western European thought, one speaker called for increased attention to Russian thinkers as they developed independently from the West. According to him, "bourgeois cosmopolitan historiography" argued that Russian thought was derivative of Western European thought. He traced the Russian revolutionary tradition back to nineteenth-century Russian thinkers, demoting the influence of German philosophy in the process.[76] For him, Russian progressivism and materialism developed in contrast to Western trends, not in coordination with them. Philosophers, he argued in a neo-Slavophile vein, needed to "destroy the myth of dependence."[77] Other speakers reacted strongly against this emphasis on Russian national tradition, particularly since it seemed to diminish the international significance of Leninism and Soviet Marxism precisely when non-Soviet socialist countries were forming.[78]

Zhdanov was among those who criticized Aleksandrov's treatment of the history and influence of Russian thinkers. Despite mentioning in his introduction the importance of Russian ideas in the development of thought in the West, Aleksandrov did not explore this connection in the body of the book. Nor did he discuss the history of Russian thought. Zhdanov pointed out that Aleksandrov seemed to imply that Marxism was primarily a regional current emerging from the West.[79]

Debates about the balance between Party militancy and scholarship and between Russocentrism and Marxist internationalism suggest that the conference was more than simply an attempt by Party leaders to dictate a Party line. The Party insisted that philosophers themselves come up with viable solutions to the problems in the field. Aleksandrov ended the meeting with a long speech admitting his mistakes. He had failed to engage bourgeois philosophers and to recognize that his textbook would be used as an ideological weapon and thus needed to be livelier, have more of a "militant spirit," and explore in greater depth the historical conditions of philosophical systems. He accepted that he had not shown the similarity of all the philosophical systems that had preceded Marx, and admitted that he had not given enough emphasis to Russian philosophy. He pledged to use "all of my propaganda and other experience to execute the tasks given to us by the Central Committee of the Party and by Zhdanov." Noting that philosophers had at their disposal institutes, departments, and publishers, Aleksandrov implored them to work together to improve their production. He ended his speech with a promise to Stalin and Zhdanov on behalf of philosophers to struggle for the "elevation of philosophical work in the country and for the organization of the wide propaganda of Marxism-Leninism."[80]

On the surface, Zhdanov's rousing speech and Aleksandrov's exercise in "self-criticism" seemed to have set the stage for renewed vigor on the philosophical front. But, on closer inspection, the results were not so clear. As one participant later recalled: "Zhdanov's speech made a strong impression on the participants in the discussion . . . as a large-scale synthesis and a global formulation somehow raising historical-philosophical methodology to a new and high level. . . . [but] the first impression was superficial and dissipated as soon as it was possible to analyze the text in its published form."[81] Indeed, in the speech approved by Stalin, Zhdanov did synthesize much of what had been said at the discussion. But he failed to outline any concrete measures for the improvement of work on the philosophical front. He rejected the one strong proposal for change, the creation of a philosophical journal. Zhdanov did not even hint that Aleksandrov would lose his post as the head of Agitprop or that another shake-up in philosophy would take place. Instead, the publication of the speeches as well as the contributions of the dozens of scholars who had not had a chance to speak at the conference itself would reveal to the public and the world the myriad problems facing Soviet philosophers. Acting as though exposing troubles would automatically lead to their dissolution, Zhdanov and Stalin had identified the weaknesses of Aleksandrov's book and the depravity of work on the philosophical front as a whole. They had not addressed the question of what to do next.

While the philosophy discussion received considerable attention and the press hoisted Zhdanov's statements into the pantheon of "brilliant" Party interventions on behalf of more stringent ideology, the tangible repercussions remained unclear. The Party had asserted its prerogative as the final interpreter of Marxist-Leninist doctrine. Agitprop would closely monitor any aspect of the teaching and publication of philosophy. At the same time, Party leaders continued to be frustrated about the situation on the philosophical front. Philosophers were having a hard time producing ideas and formulations that the Party could subsequently judge.

Over the summer of 1947 the Central Committee ordered personnel changes in the Party bureaucracy and in the leadership of the field of philosophy. In September it removed Aleksandrov and his assistants from Agitprop. Mitin and Beletskii and others who had led the attack against him, however, did not directly benefit from the move. Instead, Mikhail Suslov and Dmitrii Shepilov, neither of whom had close ties to the existing factions in philosophy or to Zhdanov, took control of Agitprop. Zhdanov did assure that he had one close ally remaining in the ideological administration: his son Yuri Zhdanov became the head of the Science Section within Agitprop, a position from which he could closely monitor and influence future academic disputes.

Aleksandrov did not remain in limbo for long. Within a week of his dismissal, the Politburo named him director of the Institute of Philosophy, a position that required him to report to the bureaucrats who had replaced him in Agitprop. But he remained in a powerful position within the field of philosophy. If Mitin and others had hoped that Stalin's criticisms of Aleksandrov's book would bring about a thorough overhaul of the philosophy hierarchy, they must have been disappointed. Instead, Agitprop was taken away from philosophers entirely and handed over to Party functionaries who had not played central roles in the philosophical discussion at all.

Aleksandrov's more dogmatic critics failed to benefit from another move by the Central Committee: somewhat surprisingly, given the tenor of the discussion, the Party created the journal *Questions of Philosophy* as a forum for philosophical work. Evidently, after the discussion, Andrei Zhdanov reported to Stalin once again that philosophers had expressed a strong desire for a journal. Stalin, who had approved Zhdanov's speech dismissing the need for the journal, now told his right-hand man that the philosophers should be allowed their own organ. But he ominously emphasized that philosophers would be held personally responsible for its contents.[82] The Central Committee appointed Kedrov, who had been one of the chief advocates of the journal and whose expertise in the philosophy of science placed him in a good position to adjudicate persistent philosophical controversies, to be the journal's chief editor. The first issue of the journal published the speeches from the discussion of Aleksandrov's book. But, beginning with the second issue, battles ensued among philosophers and within the editorial board about the journal's mandate: Kedrov saw the journal as a forum for discussion, whereas workers in Agitprop saw it as a mouthpiece for the Party, meaning that articles should emphasize unanimity, not disagreement.

Two of the new Agitprop workers, Shepilov and Yuri Zhdanov, reported to their superiors that the plan for the second issue was unacceptable: "Major themes from the fields of dialectical and historical materialism were not in the field of vision of the editors—and this is six months after the philosophy discussion." At best, they continued, the journal was "running in place"; at worst it was "taking a step backward" by trying to "settle old scores with the previous philosophical leadership."[83] Having pushed the issue with Stalin, the elder Zhdanov must have felt particularly responsible for the journal. When he heard of the conflicts surrounding its content, he called in Kedrov, who offered to change the controversial articles he had slated for the journal. Andrei Zhdanov "categorically objected" to Kedrov's acquiescence, stating, "We understand your difficulties, especially at the beginning . . . but . . . we will be lenient and therefore do not be afraid, we will not let them push you around and we will firmly support [you]." Zhdanov emphasized how important it was to get the journal out quickly, perhaps because the delays would reflect poorly on him personally.[84] The second issue did come out with controversial articles about the philosophy of science, and as a result the journal faced continual challenges from Agitprop and philosophers such as Mitin, who did not agree with Kedrov's open-minded approach to the field. By early 1949, Mitin's relentless pressure had paid off: the Central Committee removed Kedrov from his post.

The troubles in philosophy continued throughout the late Stalin period despite changes in the editorship of *Questions of Philosophy*. A memo from Yuri

Zhdanov to Suslov noted that even though two years had passed since the philosophy discussion, the philosophical front remained in crisis. The tasks given to philosophers had not been fulfilled. A new history of philosophy based on the discussion had not been written, "and judging by the way things are organized it is doubtful that it will be successful." *Questions of Philosophy* had not developed into a leading journal; philosophers were engaged in group battles. He added that the Party needed to address "in the sharpest manner" the troubles in the field, in no small part because they were having an "unfavorable effect" on "other fields of scholarship."[85]

The generally morose tone about the conditions on the philosophical front led Agitprop to a startling conclusion about philosophy in the USSR: "Marxist-Leninist philosophy had been developed and moved forward not by professional philosophers, freed from all activities besides teaching and writing philosophical books and articles, but by political figures, practical revolutionaries and scientists." Yuri Zhdanov lamented that not a single professional philosopher in the whole history of Soviet power had come up with a new thought worthy of inclusion in the classics of Marxism-Leninism or even an idea that enriched one concrete area of thought. "The majority of our professional philosophers made mistakes in their work, diverged from Marxism-Leninism. However, in posing and answering truly new questions, all of them, without exception—the incorrect and the correctors, the criticized and the criticizers—remained and remain completely and absolutely sterile."[86]

The persistent troubles in Soviet philosophy of the late Stalin period are neatly exemplified by the stumbling effort to write a multivolume history of philosophy. After the failures with volume III and then with Aleksandrov's book, Stalin once again charged philosophers with the task of writing a comprehensive history of philosophy. The Politburo decreed that the work should be done in twelve to eighteen months. But in 1949 the "due date" came and went, and the authors asked for an extension. In the middle of 1950 Agitprop complained that the book was still nowhere near completion, which was no exaggeration. The Institute of Philosophy finally published the new history of philosophy in 1957, eight years late and nearly four years after Stalin's death.

According to the Central Committee's interpretation of the "philosophical front," the problems in philosophy persisted despite Stalin's meeting with philosophers and Zhdanov's speech calling for renewed vigor in the field. It does not seem to have occurred to Party bureaucrats that the participation of the highest authorities in the land may have had a detrimental impact on the productivity of scholars. When leaders intervened on behalf of "criticism," they considerably diminished the chances that a new scholar would step forward and present work on a subject where the Party line was unclear. The ridicule heaped on outspoken philosophers such as Mitin, Aleksandrov, and

Kedrov—no matter how justified it might have been—could only discourage others from taking a chance. As a result, the field remained calcified, and any potential innovation was smothered under the weight of repetitive quotations from Marx, Engels, Lenin, and Stalin. Some subjects, including the history of philosophy, were better left untouched until the potential price for saying something incorrect had dissipated.

The discussion's impact went far beyond the field of philosophy. When Aleksandrov lost his position in Agitprop, Zhdanov lost a key ally in the ideological wing of the Party. Suslov and Shepilov, the new figures in Agitprop, did not owe their advancement to Zhdanov and could not be counted on to support him in his conflicts with Malenkov and Beria, his two main rivals in the Party. Likewise, when Kedrov's journal stumbled and had to be thoroughly reorganized, it reflected poorly on Zhdanov, who had staked much of his legitimacy on the reinvigoration of the ideology. As one historian of the political battles of the period has observed, in the aftermath of the philosophy discussion Zhdanov became more and more like a general without an army.[87] His speech at the 1947 philosophy discussions should be understood as both the peak of his power and as a crucial step toward his demise.

The end of the Second World War and the emergence of the Cold War struggle for the support of peoples around the world who remained unaligned suggested to Stalin that Soviet philosophers had to provide an ideology of victory and a blueprint for future progress that had global appeal. He recognized that ingenuity was needed from below, but that philosophers had been trained to wait for word from above. When Stalin's criticisms and Zhdanov's proclamations took the form of general statements, philosophers attempted to interpret them. But loyalty to the Party could not resolve the confusion caused by the evolution of ideological demands. A livelier tone could be adopted and the genre of the Party press could be applied to scholarship, but the deeper problem of not knowing what to say remained. Fresh ideas were hard to come by. Increasingly, the Party turned to the Russian scientific tradition as a source to advance dialectical materialism and to bolster Soviet pride in the struggle with the West. Ivan Pavlov in physiology, Ivan Michurin in biology, and Nikolai Marr in linguistics were the Party's best hope for native advancement of philosophy as a science. If dialectical materialism was the only foundation for science, then it seemed to follow that genuine development in the sciences could lead to the invigoration of dialectical materialism. During the remaining years of Stalin's life, efforts to define ideologically correct science mirrored the philosophy discussion on a structural level, with persistent infighting bubbling to the surface in grand but ultimately vacuous semipublic and semischolarly declarations of an untenable Party line. These discussions of science, however, must be viewed against the backdrop of the general crisis in ideology and the growing sense within the Party that the future of Soviet Marxism would not rest with those who called themselves philosophers.

CHAPTER 3

"THE FUTURE BELONGS TO MICHURIN"

The Agricultural Academy Session of 1948

With Soviet dirt under his fingernails and the gleam of the great communist future in his eyes, Trofim Denisovich Lysenko fancied himself a leading participant in the Great Soviet Experiment and a scientist of the people. His ideas about plant heredity sought to address devastating problems in agriculture while emphasizing the harmony between science and Marxism-Leninism.[1] Whereas Aleksandrov and other philosophers had failed to produce scholarship that both reflected and furthered the official Soviet worldview, Lysenko seemed to represent everything Andrei Zhdanov's campaign had called for. In the fall of 1947 Zhdanov had declared at the inaugural Cominform meeting in Poland that the world was divided into "two camps"—socialist and capitalist.[2] Likewise, Lysenko insisted on the distinction between "bourgeois" and "proletarian" biology.

Despite Lysenko's obvious flair for blending scientific theory and Party tenets, Zhdanov did not outwardly support him. Why would these two ardent Stalinists not see eye to eye? First, Zhdanov accepted the position of geneticists who argued that Lysenko's tactic of smearing his opponents threatened scientific productivity. He may have even believed that Lysenko's scientific ideas were wrong. Through 1948, key subordinates of Zhdanov's in Agitprop sided with the geneticists and resisted Lysenko's efforts to divide Soviet biology into opposing camps. But Stalin clearly liked Lysenko, so attacking him openly was not expedient, even for Zhdanov. Instead, he tried to limit Lysenko's influence. Second, Zhdanov resisted supporting Lysenko because it would have weakened his position vis-à-vis Stalin's other lieutenants. Ly-senko's main patron in the Central Committee—besides Stalin—was Malen-kov, who oversaw agriculture. Inasmuch as Lysenko proposed hands-on solu-tions to agricultural problems, he reported to Malenkov. But as a scientist, he fell under Zhdanov's jurisdiction as chief ideologue. Lysenko became a natural focal point for the Kremlin intrigues that pitted his two superiors against one another. While Malenkov and Zhdanov struggled for position, Lysenko's goal remained fixed: to win Stalin's open support for his ideas.

Beginning in the late 1920s Lysenko, a Ukrainian from a peasant family, amassed considerable prestige by expressing overwhelming confidence in the ability of Soviet science to revolutionize man's relationship to the natural world. His first major breakthrough came with his discovery of what he called "vernalization." By dampening or cooling winter varieties of wheat, Lysenko claimed that he could turn them into spring varieties capable of producing higher yields. He either did not know or simply ignored the fact that other scientists had worked on, and abandoned, this idea. Soon vernalization be-came Lysenko's recipe for solving a full range of agricultural problems.[3] That there was no experimental base for his assertions did not seem to deter him or his growing list of followers in the agricultural establishment. Revolutionary zeal, not measured caution, was the order of the day.

Though he never joined the Party, Lysenko quickly learned to put his novel scientific ideas in the context of Marxist-Leninist thought. With the help of his ideological sidekick I. I. Prezent and the philosopher Mark Mitin, Lysenko came to view biology as the center of a great "ideological battle" that pitted "idealists" against "materialists," categories borrowed—very loosely—from Lenin. The idealists—following in the tradition of the Germans Gregor Men-del and August Weismann and the American T. H. Morgan—isolated the key to evolution in chromosomes. The materialists—a label Lysenko assigned to himself—believed that environment played the central, if not sole, role in evolution. The materialists also rejected outright the attempt to identify he-reditary material in cells.

Lysenko's work had some striking similarities to the ideas of Jean-Baptiste Lamarck, a nineteenth-century French proponent of the inheritance of ac-quired characteristics. But Lysenko emphasized that he was carrying on a spe-cifically Soviet tradition in agronomy. Instead of looking to Weismann and Morgan or even Lamarck for guidance, Lysenko elevated the Russian Ivan Michurin to heroic status. Michurin, a semiliterate plant breeder, believed that his practical experience was far more valuable than the skeptical positions of theoretical scientists. Before the Revolution, scientists attuned to the math-ematical regularity of genetics had dismissed Michurin's miraculous claims about the unlimited potential of cross-fertilization. Though geneticists persis-tently challenged his ideas in the 1920s, they could do nothing about the public relations campaign that the Bolsheviks launched in his favor.[4]

Lysenko also made sure to contrast his humble origins with the "bourgeois" trappings and demeanor of his opponents. With the help of journalists eager to show a man of the people actively discovering new ways to improve crops, Lysenko, like his trailblazer Michurin, became a cultural hero. The press followed and praised Lysenko's work to such an extent that his fellow scientists usually did not find it expedient to publicly criticize him. Throughout the 1930s Lysenko gathered institutional and political clout as well, especially in the area of agriculture, where Central Committee bureaucrats clearly favored him. He became a full member of the Ukrainian and the Soviet Academies of Sciences, secured for himself control over the Lenin All-Union Agricultural Academy, and even was elected to the Supreme Soviet of the USSR, nominally the highest legislative organ in the country. As his authority grew, so did his vicious denunciations of those who challenged his ideas. Lysenko's greatest political coup of the period occurred in 1935 when Stalin responded to a speech he gave at the Kremlin by calling out from the audience: "Bravo, Comrade Lysenko, Bravo."[5] As with so much of what Stalin said, reporters repeated the quotation, which further added to Lysenko's status.

Lysenko's conflicts with geneticists came to a head in a discussion at the Marx-Engels-Lenin Institute in 1939. A number of scientists had written to Zhdanov to complain about Lysenko's vitriolic language and attempts to gain even more bureaucratic control of biological research. Zhdanov appeared to be impressed with their arguments: he underlined portions of their letter that mentioned Lysenko's abuse of his administrative power, his misappropriation of Marxism, and his erroneous experiments and theories. But the Central Committee Secretariat left the responsibility of adjudicating the dispute to philosophers, under the direction of Mitin.[6] It is important to emphasize that neither the geneticists nor Lysenko were clearly "more Marxist" in any significant sense. Marx and even Engels had not written enough about the natural sciences for anyone to reach definitive conclusions about what Marx or Engels had believed about specific disciplines. But Lysenko's ideas did sit well with the Soviet understanding of science and its importance in building communism. Lysenko promised dramatic improvements in agricultural production to a state that periodically suffered from famine. To great rhetorical disadvantage, the geneticists studied evolution by examining fruit flies, *Drosophila*, and had a harder time showing the practical applications of their work. Still, the results of the 1939 discussion were ambiguous: Mitin and his cohort accepted the geneticists' theoretical claims and denounced Lysenko's bureaucratic abuse, but they clearly endorsed Lysenko's emphasis on practical rewards and his professed adherence to a Soviet scientific tradition.[7]

In the immediate postwar period, geneticists made considerable strides within the Party bureaucracy, which suggested that the compromise of 1939 might be reexamined in their favor. Furthermore, Mitin and his cohort had lost considerable influence after Stalin lambasted the three-volume history of philosophy. With his colleagues in philosophy in retreat, Lysenko could not gain any advantage, even after Zhdanov's attacks on ideological mistakes in literature and the arts. In 1945 Anton Zhebrak, who had studied at the geneticist T. H. Morgan's laboratory at Columbia University in the early 1930s, became an official in the Central Committee's Science Section. He immediately worked to diminish Lysenko's influence. In 1946 and 1947 the Science Section, headed by Sergei Suvorov, sent a number of letters to Andrei Zhdanov complaining about Lysenko. Rather than attacking Lysenko's ideas, Suvorov concentrated on Lysenko's tight control over the Agricultural Academy. At first, even the agricultural ministers agreed with Suvorov that the academy needed an influx of new members who would improve its practical work, broaden its mandate beyond Lysenko's pet projects, and ensure that its leaders were reputable scientific workers. When Lysenko threatened to resign over the issue, Zhdanov and the Central Committee pushed slowly forward, calling for new elections to the academy to take place.[8]

Lysenko and his colleagues fought back, writing to Zhdanov about scientists at Moscow State University (MGU) and in the Academy of Sciences who, they believed, were propagating "reactionary" and "proto-fascist" ideas in the Western press. Their main targets were Zhebrak and Nikolai Dubinin, who had published articles in the American journal *Science* in which they surveyed the accomplishments of genetics research in Russia while distancing their work and Soviet science in general from the negative impression made by Lysenko.[9] But Lysenko failed to win Zhdanov's endorsement, which at that point appeared to be the key to winning any ideological battle in the Soviet Union.

In 1946 and the first half of 1947 Zhdanov was Stalin's heir apparent. He chaired the Central Committee's Secretariat and ran the Politburo whenever Stalin was away from Moscow, which after the war sometimes stretched to months at a time. While Zhdanov oversaw the Party's day-to-day operations, Stalin clearly kept abreast of major policy moves, including the campaign to reinvigorate ideology. Zhdanov's speeches and the Central Committee's decrees forced artists, writers, musicians, and scholars to recognize that the Party would judge their work by the degree to which it adhered to Marxist-Leninist principles and the demands of patriotism and loyalty. His influence extended deep into the Party and state bureaucracy as well. He placed his protégés in key positions at the expense of his main rivals among Stalin's lieutenants, Malenkov and Lavrenty Beria. In May 1946 Stalin removed Malenkov from the Party Secretariat. Though Malenkov officially remained a member of the Politburo, he was clearly being punished—possibly at Zhdanov's instigation—for his poor organization of the aviation industry and his handling of reparations from Germany.[10] A. A. Kuznetsov, who had worked with Zhdanov in Leningrad, took Malenkov's place as active member of the Secretariat. Beria's setback was relatively minor. He remained a key figure in Stalin's entourage and held the strategic position of overseeing the Soviet atomic bomb project. Still, Stalin eroded some of Beria's influence by placing Kuznetsov in control of the Party's supervision of the security ministries.[11]

Zhdanov's rivalry with Malenkov had its roots in the 1930s, when they clashed over the organization of the Party apparatus. Malenkov, whose main influence emanated from his work with the industrial and agricultural ministries, wanted the Party to take a more active role in organizing the economy. Zhdanov, in contrast, believed the Central Committee needed to concentrate on agitation and political work.[12] Zhdanov had joined the Party prior to the Revolution, when the Bolsheviks' accession to political power had still been only a remote possibility. The most important revolutionary activities at that point involved studying and propagating Marxism. Neither Beria nor Malenkov, who joined the Party in 1917 and 1920 respectively, showed much of an inclination to debate the finer points of theory. After the Second World War, Stalin endorsed the idea of the Party as a vigorous proponent of ideological purity, and as a result Zhdanov emerged as the strongest of his lieutenants.

In early 1947, however, minor restructuring in the Party apparatus suggested that Zhdanov's power was beginning to wane in ways that would benefit Lysenko. I. A. Benediktov, who supported Lysenko's views, became the minister of an enlarged Ministry of Agriculture. Around the same time Beria assisted Malenkov in reestablishing himself as the assistant chairman of the Council of Ministers and in the Party apparatus. Malenkov quickly became involved in agricultural matters. In March, Benediktov and two like-minded colleagues wrote to Zhdanov and Malenkov complaining about a genetics conference at MGU and labeling the head of the Genetics Department there a "formal geneticist" who had made "eugenics mistakes." Again Suvorov dismissed the Lysenkoist language, informing Zhdanov that "the genetics conference . . . was entirely useful and the attempt of Comrades Benediktov, Lobanov, and Kozlov to discredit it is unjust and based on one-sided information."[13]

In April, Suvorov once again raised the possibility of holding new elections to the Agricultural Academy. This time the Central Committee passed a resolution requiring Lysenko to deliver an address to the Orgburo about the state of the academy. Lysenko reiterated his views about the differences between the "metaphysical" and "bourgeois" ideology of his opponents and his own dialectical materialist biology. He complained that the field had come under the influence of foreign science, a xenophobic theme that jibed with the broader ideological campaign. In a subtle way, he also expressed frustration about Suvorov's actions, which prevented him from winning over the Party's highest echelons: "I am literally tormented by the fact that up to this point I have not been able to bring the state of biological and agricultural sciences in this country to the attention of the government and Party." Lysenko must have recognized that Suvorov was pointedly keeping him and his biological views outside the confines of the ideological campaign.

The results of the June 1947 discussion of philosophy, however, exposed Zhdanov's vulnerability. He had lost key allies in Agitprop, which served as the main engine of the political work that he had insisted was the Central Committee's primary function. Beria and Malenkov, who at this point had little direct interest in the Party's ideological battles, may have recognized in them an opportunity to embarrass Zhdanov. After all, Aleksandrov had written and published his *History of Western European Philosophy* during Zhdanov's watch. The recent exposure of errors among intellectuals suggested that the Party, and by extension Zhdanov, had done a poor job of anticipating and preventing deviant ideas.

A number of other events in the summer of 1947 bolstered Lysenko's position. First, Stalin and Zhdanov established so-called Honor Courts to attack unpatriotic behavior and subservience to the West. In the fall Zhebrak, who had been the most influential geneticist in the Party, became a victim of one of these courts. Mitin, Lysenko's old ally among philosophers, had recently been appointed the science editor at the *Literary Gazette*, and he used his position to lead the charge.[14] But despite these advantageous moves, as long as Suvorov and Zhdanov were insistent on new elections to the Agricultural Academy, Lysenko remained vulnerable. In short, the Party's position on Lysenko's controversial theories and tactics remained ambiguous. Malenkov and the Ministry of Agriculture supported Lysenko. Zhdanov and the Science Section of the Central Committee remained more skeptical. It was up to Stalin to decide the issue.

By the fall of 1947, Zhdanov had more to worry about than the situation in biology or his political struggles with Malenkov. Though he was only fifty-one years old, a heavy drinking habit and long nights working as a member of the Politburo had taken their toll. In October, hypertension and exhaustion forced him to leave Moscow for Sochi, a resort town on the Black Sea. After one month, his doctors reported to Stalin that his condition remained serious and asked that his treatment be allowed to last until early December. The Politburo granted the request.[15]

With Zhdanov away from the capital for most of the fall, the restructured Agitprop headed by Mikhael Suslov and Dmitrii Shepilov—neither of whom were close allies of Zhdanov—took over the administration of the ideological front and monitoring the battles in biology. Zhdanov was not completely out of the loop. His twenty-eight-year-old son, Yuri, took over Suvorov's position as the head of Agitprop's Science Section. While nepotism was clearly a factor in his gaining such a prestigious appointment, Yuri Zhdanov also had the necessary educational background for the job. He had followed up his study of chemistry at MGU with graduate work in the philosophy of science with B. M. Kedrov, the editor of the new journal *Questions of Philosophy*.[16] His education, however, could not make up for his inexperience in politics and in the workings of the Central Committee. At the beginning of his tenure, the Science Section and Agitprop often failed to deliver clear messages about what constituted ideologically correct science.

In October 1947, Lysenko took advantage of Zhdanov's absence and the confusion in Agitprop's Science Section. In late 1946 Stalin had personally provided him seeds of grain that were planted in Odessa, Omsk, and outside Moscow. Now he wrote to Stalin directly, bypassing the disorganized Agitprop and Zhdanov's Secretariat.[17] The first section of his letter concentrated almost exclusively on the ways in which he was trying to further the agricultural output of the USSR. He emphasized that he was working within the framework of Stalin's previous advice and that he needed Stalin's support to continue to make breakthroughs.[18]

Having established his credentials as a practical scientist and his potential contributions to Soviet society, Lysenko turned to the question of his scientific rivals. He began with general "theoretical" concerns.

> Dear Comrade STALIN! I believe it is also necessary briefly to discuss theoretical biological conceptions, which originate from the aforementioned plan for practical work and from all of my other work as well. I champion the position that the primary reason for the new growth in strains of plants and animals, as well as the consolidation of these strains, is their changing nature in new environments.[19]

Lysenko pointed out that the "Mendelian-Morganists" did not accept the notion that the environment could affect the inheritability of characteristics and relied instead on random natural mutation. He also emphasized fundamental differences between "our Michurin-based genetics" and the "dangerous and dishonest . . . Mendelian-Morganist genetics," which was being taught to students throughout the country. He equated "Neo-Weismannism" with "bourgeois metaphysics," which sought to destroy the practical work of Michurinist science. According to Lysenko, his opponents created a gap between scientific work and its practical application on the farms of the USSR. In a sentence that was impressively dogmatic even by the standards of the time, Lysenko added:

> I dare state that Mendelism-Morganism, Weismannist neo-Darwinism is a bourgeois metaphysical science of living bodies, of living nature developed in Western capitalist countries not for agricultural purposes but for reactionary eugenics, racism and other purposes.[20]

By contrast, he continued, Michurinism emphasized agriculture, and was supported only in the USSR. It was a young science but, according to Lysenko, it was fundamentally sound. He asked Stalin to support the teaching of Michurinism and thereby revolutionize the education of Soviet agronomists, biologists, and plant breeders. He ended the letter with a final bit of flattery:

> I implore you, Comrade Stalin, to help with this good and necessary work for our agriculture. Dear Joseph Vissarionovich! Thank you for [supporting] science and [showing] your concern, which was conveyed to me during our conversation at the end of last year about branched wheat. While studying branched wheat in detail I understood much that is new and good. I will fight to make up for lost time and try to be at least somewhat useful in the broader project, that is in the movement of our wonderful homeland to an abundance of food products and in the movement toward communism.[21]

Three days later, Stalin responded with a personal letter to Lysenko expressing support for his work with the wheat varieties and hybridization and suggesting that they discuss various attempts to develop new crop varieties in the near future in Moscow. "Regarding the theoretical tenets in biology," Stalin continued, "I think that Michurin's tenet is the only scientific tenet. The Weismannists and their followers, who deny the inheritability of acquired characteristics, do not pay attention to what has been disseminated about them for a long time." Stalin then assured Lysenko that "the future belongs to Michurin."[22] Such a strong statement from Stalin was precisely what Lysenko was hoping for. Not only were his experiments given official endorsement, but he had successfully outflanked Zhdanov and others in the Central Committee apparatus who had stood in his way. It simply remained to be seen when and how the Michurinist future would be declared officially.

Stalin, however, played his cards carefully. A few weeks later he distributed Lysenko's letter to all the members and candidate members of the Politburo and the secretaries of the Central Committee. But, significantly, Stalin did not share his response. Thus, it is possible that the highest Party officials were left in the dark about the leader's position on the theoretical battle. To them he wrote that he wanted them to familiarize themselves with this subject of "fundamental importance and urgency." "In due course," he added, "the questions raised in the letter will be discussed in the Politburo."[23]

With Stalin's views still under wraps, the Science and the Agricultural sections of the Central Committee, the Ministry of Agriculture, and the Academy of Sciences set about preparing for the Politburo discussion. In November 1947, a large meeting at MGU held to discuss genetics and evolution ended with a rout of the Lysenkoists, who refused to speak, despite invitations to defend their scientific views.[24] In December 1947 the Biology Division of the Academy of Sciences held a closed meeting to discuss species competition. Lysenko and Ivan Ivanovich Shmal'gauzen, the chair of the Department of Darwinism at MGU, openly clashed. Although L. A. Orbeli, the head of the academy's Biology Division, was reluctant to support Lysenko's opponents in the press, he was determined to keep Lysenko's influence in the Academy of

Sciences to a minimum.[25] The December meeting was intended to dismiss Lysenko's most recent "theoretical breakthrough"—that no competition for survival occurred within species in nature.[26] Again, the Central Committee did not take sides in the argument.[27] Some biologists were eager to put Lysenko on the defensive, but Andrei Zhdanov, now back in Moscow, refrained from offering them explicit support. Though Stalin's views may have remained unknown to Zhdanov at this time, it was clear that confronting Lysenko directly could be dangerous considering that the planned "Politburo discussion" had yet to take place.

In January 1948 the struggle over the Agricultural Academy heated up. A committee chaired by Agitprop deputy chief Shepilov reported to Party secretary Suslov that a list of candidates had been drafted and that new academicians could be elected at a meeting in February. As usual, Lysenko offered his own dissenting report and refused to endorse the list of candidates. The Secretariat, powerless to force changes on Lysenko before the planned Politburo discussion, delayed elections once again.[28]

Lysenko's position in the Academy of Sciences and among university scientists was clearly not strong. But he used a strategy he had developed in the 1930s: whenever he was on shaky scientific ground he extended the debates beyond the confines of the academic community. Mitin took up Lysenko's cause and used the Literary Gazette as a forum for arguing that the debates among biologists were of central importance for everyone in the Soviet intelligentsia. In this context, Lysenko's insistence on ideological purity and Soviet patriotism carried more weight. Biologists may have realized that there were numerous examples in nature of intraspecies competition, but if the discussion had a broad impact on Soviet thought, then the Party would have the final say. The scholarly press was more evenhanded than the Literary Gazette. Kedrov allowed Shmal'gauzen to present his anti-Lysenkoist views on the interaction between Soviet philosophy and biology in Questions of Philosophy.[29]

Throughout the spring of 1948, individual scientists, sensing that a decree on the "theoretical disputes" in biology was imminent, appealed to the Central Committee in an effort to sway the Party's opinions. Shmal'gauzen took the lead. In early January he wrote two letters to Andrei Zhdanov to complain about the Literary Gazette's coverage of the intraspecies debate.[30] He feared that a "black wave" in the press had been unleashed simply because the biology faculty at MGU had "openly criticized Academician Lysenko's thesis about the absence of intraspecific competition."[31] The Science Section summoned Shmal'gauzen to discuss the question, but he did not receive tangible support. When he asked to publish an article on intraspecific competition in Culture and Life, the organ of Agitation and Propaganda, Shepilov rejected the request.[32]

Lysenko faced ridicule from yet another, more surprising, angle. In the late fall, when Stalin circulated Lysenko's letter to the Politburo and Secretariat, he also asked that a copy be sent to Nikolai Tsitsin, the vice president of the Agricultural Academy. Tsitsin—who had supported Lysenko over the years—now wrote a thirty-page letter to Stalin and Party secretaries Malenkov, Zhdanov, and Kuznetsov, complaining that Lysenko's ideas about branched wheat had no scientific merit. Tsitsin placed blame for the serious problems in biology squarely with Lysenko and his attempts to divide Soviet science into "Michurinist Darwinism" and "reactionary, bourgeois" camps. Tsitsin complained that, "in the last 20 years [Lysenko's theories] have not produced one decent variety, despite the great number of promises and loud assurances." But Tsitsin's letter, like Shmal'gauzen's, was quickly shelved.[33]

The arguments presented by Shmal'gauzen, Tsitsin, and others may have compelled Yuri Zhdanov to join the offensive against Lysenko. He did not start with an all-out attack, nor did he follow Suvorov's example of quiet, but persistent, complaints about Lysenko whenever the opportunity presented itself. In February, Yuri Zhdanov went right to the top with his views. He wrote a long memo to his father, Malenkov, and Stalin about Lysenko's hindering of the development of "kok-sogyz," a rubber plant variety: "From the very beginning, instead of objectively studying the new breed . . . he [Lysenko] has created a poisonous atmosphere of hostility and mistrust." The younger Zhdanov saw Lysenko's interference with research on kok-sogyz as an example of how his leadership had hindered practical work and negatively influenced potentially valuable science.[34]

Lysenko countered in a letter to Malenkov that work on kok-sogyz would "do nothing less than legitimate Mendelism-Morganism in our country." There is no record of any response from Stalin on the question, but it is clear that at this point Malenkov was squarely in Lysenko's camp.[35] Still, Yuri Zhdanov must have been confident that he could win Stalin over and interpreted his silence as an invitation to develop further his views about the situation in biology. This time he chose a more open forum.

On April 10, Yuri Zhdanov gave a long speech to a group of propaganda workers in which he strongly attacked Lysenko for attempting to monopolize biological research, for promoting "unscientific" concepts such as the denial of intraspecies competition, and for using "unscientific" tactics. He was careful to note that the opinions expressed in the speech were his own, and that he was not speaking as the head of the Science Section or on behalf of the Central Committee. He recognized that there was a struggle going on between two biological schools, but he insisted that both could be considered Soviet and neither should be called "bourgeois." He criticized Lysenko's habit of calling everyone who disagreed with him "bourgeois" or "anti-Soviet" and Mitin for using the Literary Gazette as a forum for Lysenko's views without giving his opponents the opportunity to respond.[36]

Yuri Zhdanov's talk was greeted by the anti-Lysenkoists as a very positive sign. As Andrei Zhdanov's son, he may have been one of only a few people in the country with the political clout to openly attack someone of Lysenko's stature without fear of immediate retribution. Kedrov, who was under pressure for not supporting Lysenko in the journal *Questions of Philosophy*, addressed the significance of the talk before the Party organization of the Institute of Philosophy. While acknowledging that Zhdanov was not speaking in his official capacity as the head of the Science Section, Kedrov emphasized that the speech was prepared in advance and implied that it had been approved at some higher level. The Central Committee's position still remained unclear but, in Kedrov's opinion, the speech had shown that the biology taking place in the Soviet Union should not be divided into "bourgeois" and "proletarian" approaches. It was possible to be critical of Lysenko without being anti-Soviet.[37]

Lysenko responded quickly. He had listened to Yuri Zhdanov's speech while sitting in Mitin's office, which was located near the lecture hall, and he was infuriated. He understood the potential damage that such a strong statement coming from the head of the Central Committee's Science Section could cause. His subsequent actions make it clear that he wanted Stalin to get involved more directly in the dispute. On April 17, he wrote to Stalin complaining about the speech, suggesting that the "anti-Michurinist slander" could have negative practical consequences:

> [the] falsehoods of the anti-Michurinist neo-Darwinists will have much greater effect in the regions, both among scientific personnel and among agronomists and officials of practical farming, thereby strongly hindering the scientists under my direction from applying their results in practical farming. I turn to you, therefore, with a request that is very important to me: to help, if you consider it desirable to do so, in this matter that seems to me very serious for our agricultural science and biology.[38]

When this letter did not get a response from Stalin, Lysenko continued to press the matter with Malenkov. It is likely that he discussed Yuri Zhdanov's speech with Malenkov before sending Malenkov a long memo responding to it. He also wrote to the minister of agriculture offering his resignation as the head of the Agricultural Academy. Lysenko understood that this post was part of the Politburo's nomenklatura and therefore his resignation required a decision by that body, which was chaired by Stalin himself.[39] In this way, he could be sure that the discrepancy between Stalin's views and those expressed by Yuri Zhdanov in his speech would become evident. As Lysenko had hoped, Stalin became directly involved.

On May 28, 1948, Stalin called Malenkov, Suslov, Shepilov, both Zhdanovs, and other Party leaders into his Kremlin office.[40] A number of accounts of the meeting are available, and while they differ on some points, there is agreement that Stalin was very angry about Yuri Zhdanov's speech. He reprimanded the young apparatchik for expressing his personal view about the situation in biology when the Central Committee's position had not yet been clarified. "The Central Committee," Stalin insisted, "can have its own position on scientific questions."[41] In December Stalin had said that the Politburo would discuss theoretical questions of biology, and evidently the time had come. Again, as was the case with philosophy the year before, Stalin seemed determined to establish a Party line when he perceived that the apparatus itself had not managed the issues well enough. Shepilov later reported on what took place at the meeting: "Pipe in hand and puffing on it frequently, Stalin paced the room from end to end, repeating practically the same phrases over and over: 'How did anyone dare insult Comrade Lysenko?' 'Who dared raise his hand to vilify Comrade Lysenko?' "[42] According to another account, Stalin angrily reprimanded Yuri Zhdanov for attempting to "crush and destroy" Lysenko. The leader went on to remind his silent audience that Lysenko had recognized Michurin as a great scientist when others had called him "a tormenting, provincial crank and amateur." He admitted that Lysenko had "some shortcomings and made personal and scientific mistakes that make him worthy of criticism." Nevertheless, "greasing the wheels for all these Zhebrak types" was wrong.[43] Stalin then established a committee, headed by Andrei Zhdanov and Malenkov, to "clarify the facts" surrounding Yuri Zhdanov's lecture and the situation in the biological sciences. Stalin believed that Yuri Zhdanov had been mistaken in his speech and that there were indeed two trends in biology, one "based on mysticism" and the other "materialist." Stalin ordered that "a Marxist in biology make a report," that the Central Committee pass a resolution on the matter, and that an article appear in *Pravda* clarifying the Party's position.[44]

A few days after the Kremlin meeting, Lysenko presented to Malenkov and Stalin excerpts of Yuri Zhdanov's speech with his point by point rebuttals.[45] Stalin read the report carefully, making notes in the margins of the text. The subject was not entirely unfamiliar to Stalin, who in 1906 had written an essay in which he had expressed sympathy for Lamarck as a revolutionary scientist.[46] His interest was piqued when Lysenko wrote that the Mendelian-Morganists denied the influence of the environment on the formation of breeds. Stalin corrected him: "That's not the issue. The Weismannists-Morganists also accept the effect of the environment. Their divergence from the Michurinists lies in their denial of the hereditary transmittance of the change." Even when the geneticists accepted the influence of the environment on heredity, they believed it was not controllable. According to Stalin, the Michurinists "consider the effects to be regular and understandable, and within man's ability to control." Stalin's comments show that he was engaged in the actual substance of the debate and felt himself enough of an expert to school Lysenko.[47] He clearly believed that he was supporting the correct scientific position.

Stalin's actions also had political repercussions. While Malenkov and Andrei Zhdanov were cochairmen of the Politburo committee in charge of organizing Lysenko's formal Party endorsement, the episode marked a clear vic-

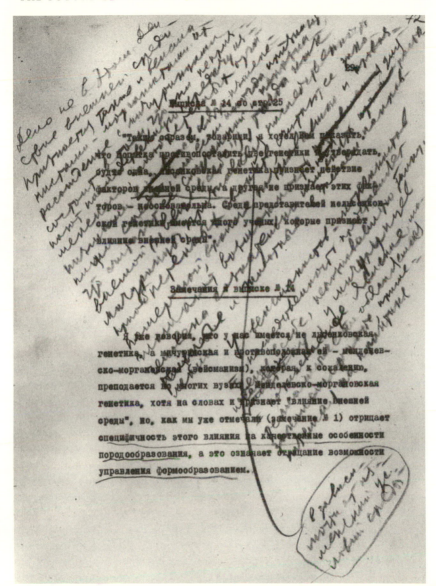

Figure 4. In the late spring of 1948 Trofim Lysenko sent Stalin his typewritten comments on Yuri Zhdanov's speech to propaganda workers. Here Stalin remarks in pencil: "That's not the issue. The Weismannists-Morganists also accept the effect of the environment. . . ." Courtesy of RGASPI.

tory for Malenkov. He had little to lose: agriculture represented only a small portion of his bureaucratic strength. But Zhdanov's power rested precisely in his dominance of ideology. To use a chess metaphor, Malenkov was willing to risk losing a knight (agriculture) to capture Zhdanov's queen (Party ideology.) As the May meeting indicated, Yuri Zhdanov's talk and Lysenko's response forced the Central Committee to make a decision regarding biology. Given Stalin's endorsement of Michurinism and acceptance of Lysenko's portrayal of two opposing camps in biology, the decision could only prove unfavorable to the geneticists, their patrons in the ideological establishment, and to Andrei Zhdanov himself. It is even possible that Suslov and Shepilov (both of whom were close to Malenkov, though subordinated to Zhdanov) were willing to allow Yuri Zhdanov to give the talk in the first place because they understood it would precipitate Stalin's intervention into the matter and lead to a clash between the elder Zhdanov and Stalin.

One of Andrei Zhdanov's last acts as a secretary of the Central Committee was to work with Malenkov on a long report titled "On the Situation in Soviet Biology." A number of drafts of this report, evidently written by Suslov and Shepilov, appear in Zhdanov's personal archive. Zhdanov edited versions of the draft himself, removing specific references to his son's erroneous attempts to reconcile the two trends in biology. The implication that the substance of the speech had been wrong remained. The Mendelian-Morganist trend was declared alien to Soviet science. No reconciliation between Soviet biology and "bourgeois biology" was possible. Michurinism "dealt a crippling blow to bourgeois biological science." On July 10 a draft with both Andrei Zhdanov and Malenkov listed as authors was sent to Stalin and distributed to all the members of the Politburo. It called for the strengthening and development of Michurinism, the condemnation and dismissal of Mendelian-Morganism as reactionary and unscientific, and the restructuring of scientific institutes, journals, publications, and educational establishments so that the Michurinist trend could prevail in Soviet biology. Soviet scientists had to accept that only by struggling against Mendelian-Morganism could they fulfill the demands of the Party. Finally, the report declared that "the discussion of this question is considered finished."[48]

Certainly in the Central Committee the time for discussion was over. On the same day the report was finalized, Yuri Zhdanov wrote a repentant letter to Stalin stating:

> The very organization of [my] report was wrong. I obviously failed to take into account my new position as an official of the Central Committee apparatus, underestimated my responsibility, and did not realize that my presentation would be taken as the official stand of the Central Committee. . . . My sharp and public criticism of Academician Lysenko was wrong. Academician Lysenko is now the recognized leader of the Michurinist school in biology, he has defended Michurin and his doc-

trine from attacks by bourgeois geneticists, and he has himself done much for science and our farming practice.[49]

Although Yuri Zhdanov now recognized the establishment of the new Party line in biology, he remained critical of Lysenko, even in his letter to Stalin:

> I disagree with some of Academician Lysenko's theoretical propositions (the denial of intraspecies struggle and mutual aid, underestimation of the internal specific character of the organism); I believe he still makes poor use of the treasure-trove of the Michurinist doctrine (which is why Lysenko has failed to develop any substantial agricultural plant varieties), and I consider that he gives our agricultural science weak leadership. The Agricultural Academy, which he leads, functions at far from full capacity.[50]

Yuri Zhdanov had clearly backed away from his support of genetics, but by criticizing Lysenko on some points, he was perhaps hoping to keep Lysenko from gaining a total monopoly in agricultural science.

During July Malenkov also oversaw the reorganization of the Central Committee. Agitprop went from an "administration" reporting to Zhdanov to a "section" subordinated to Suslov, but led by Shepilov. A number of Zhdanov's remaining protégés lost their positions, as Malenkov took over as chairman of the Secretariat and Orgburo.[51] At the same time, Andrei Zhdanov's health was quickly deteriorating. On July 5, his doctors reported that "his heart condition" had "deteriorated to the point where even regular motion causes shortness of breath" and that he had lost "feeling in the right hand and the right side of the face." They recommended a two-month leave immediately, which the Politburo granted on July 6, 1948.[52] The political battle between Zhdanov and Malenkov was essentially over.

On July 15, when Andrei Zhdanov was already away from Moscow, the Politburo passed a resolution proposing that Lysenko give a speech, to be published in the press, "on the situation in Soviet biology for discussion at the July session of the Agricultural Academy." Stalin made one significant change to the draft resolution before the Politburo accepted it. Andrei Zhdanov's effort to shield his son from direct ridicule had failed: the final resolution emphasized that Yuri Zhdanov's speech was the central impetus for Lysenko's rebuttal.[53] The Politburo resolution was immediately forwarded to Lysenko so that he could prepare his speech for the July meeting.[54]

If Stalin had published the Central Committee's draft resolution on the situation in Soviet biology, everyone in the country would have recognized that Lysenko's position had been officially sanctioned and that the discussion was officially closed. Instead, Stalin chose a meeting of the Agricultural

Academy as the platform for making Lysenko's victory known to the public. He also chose to conceal the role he had played in determining the outcome of the debate. It is likely that Stalin recognized a tension between the scientists' and the Party's claims to authority in determining scientific truth. Ideally, the new line in biology would appear as the result of a purely scientific discussion. As opposed to literature and the arts, science was supposed to have a universal, objective character and therefore, in principle, could not be declared either right or wrong simply by Party decree. Stalin was obviously comfortable with the Party publicly taking the lead in philosophical disputes, in part because philosophy was considered a Party-minded science. But a Central Committee resolution determining whose theories were correct was evidently not deemed an appropriate means of publicly resolving conflicts in the natural sciences. Instead, scientists gathered to discuss the content of Soviet science; by definition, if the science was Soviet, it would work hand in hand with Party doctrine. The public meeting was a way to give Lysenko's views a veneer of objectivity.

On July 23, eight days after the Politburo decision, Lysenko sent Stalin a copy of his speech along with the following cover letter:

Comrade J. V. Stalin
Dear Joseph Vissarionovich!

I urge you to look at my speech, "On the Situation in Soviet Biological Science," which is to be delivered for discussion at the July session of the Lenin All-Union Academy of Agricultural Sciences. I tried truthfully to set forth the situation, as well as possible from a scientific perspective. I formally avoided c[omrade] Yuri Zhdanov's talk, but in fact the content of my speech in many ways is a response to his incorrect address, which has become widely known.

I would be pleased and fortunate to receive your comments.

President of the Lenin All-Union
Academy of Agricultural Sciences

Academician T. Lysenko.[55]

By July 27, Stalin had read and provided extensive comments on Lysenko's speech. At ten o'clock that evening Lysenko met with Stalin and Malenkov at the Kremlin. They discussed the speech in private for approximately one hour.[56]

The thrust of Stalin's comments indicate that he was not comfortable with Lysenko's emphasis on the class nature of science or his attempt to equate the conflict in biology with class conflict.[57] When Lysenko wrote that "any science is class-oriented by nature," Stalin added in the margins: "Ha-Ha-Ha!!! And what about Mathematics? And Darwinism?" Some fields were clearly not class based. In a section of the speech titled "The Fundamentals of Bourgeois Biology Are False" Lysenko declared: "by its nature, the modern capitalist system cannot tolerate a true depiction of natural development" and "there is no genuine science in that [bourgeois] society." He also noted that "bourgeois genetics is 'a science' that is not a section of natural science, it is rather a metaphysical and idealistic science engendered by the ruling class." After Stalin's editing, this whole section was removed. The de-emphasis on class was also evident in other editorial changes: nine times Stalin either deleted the word "bourgeois" or replaced it with "idealist" or "reactionary."[58]

Stalin also removed a passage in which Lysenko referred to Stalin's "Anarchism or Socialism" as a guide for Soviet biologists in examining questions of species formation.[59] Evidently, even at the height of the cult of personality, Stalin did not think references to his own work were appropriate in Lysenko's speech. Finally, he removed the word "Soviet" from the title of Lysenko's speech: "On the Situation in Soviet Biological Science" became "On the Situation in Biological Science." This suggests that Stalin understood that the adjective "Soviet" could weaken the universal, scientific scope of Lysenko's claims. Stalin also wanted Lysenko himself to prove his analysis was right, not to rely on the Party's backing. Historian Kirill Rossianov's conclusion from his analysis of Stalin's editing of Lysenko's speech is worth quoting at length:

> Lysenko's arguments for the class nature of science, which could have provided a basis for the Party's interference in science, were definitely rejected by Stalin. Stalin also toned down the "political" dimension of Lysenko's language and made many places in his talk sound more "objective." The remarks on "mathematics" and "Darwinism" made by Stalin in the margins of the manuscript implied his belief in the universal character of scientific knowledge.[60]

Again the evidence indicates that Stalin intended the August session of the Agricultural Academy to reach a "scientific" conclusion to the debates in biology. Ironically, in the very process of intervening in and dictating the outcome of the discussion, Stalin accepted that the Party's power in science should at least appear to be subordinate to the power of scientific truths that were hardened by discussion and debate.

When compared with the philosophy meeting of the summer before, the meeting in biology was initially much humbler in scope and setting, and the Party's role was much less clear to the participants. Whereas Andrei Zhdanov had presided over the meeting of philosophers, the meeting of biologists had no such obvious Central Committee presence. The philosophers had met at the Central Committee building. The biologists met at the House of Scientists. Also, the biology meeting began as an in-house affair for the Agricultural Academy. It was not even clear to what extent the state bureaucracy would participate. While the Ministry of Agriculture was strongly represented at the meeting, the Ministry of Education, the Academy of Sciences, and the universities were not. It was clear that Lysenko had strengthened his hold on the Agricultural Academy, but there was no outward indication that this would be the ultimate battle between what Lysenko insistently labeled the "two opposing camps" in biology.[61]

Still, this was obviously not an ordinary meeting of the Agricultural Academy. On the eve of the session, the Council of Ministers scrapped the academy's election planned for September and simply appointed a new group of full academy members consisting exclusively of Lysenko supporters. News of the session and the list of the new members were printed in *Pravda*, giving the meeting an added aura of importance.[62]

The new appointments, however, were not an obvious sign of broader support for Lysenko in the government and Party, nor did they mean that he had succeeded in concluding the whole discussion in his favor. He had simply gained untrammeled control over the institution where he had been the president for ten years. Those who were against Lysenko still held out hope that the Party had not assured his victory elsewhere. After all, in April Yuri Zhdanov had publicly criticized Lysenko. No formal statement from the Central Committee repudiating that stand had been circulated. If word of the elder Zhdanov's illness and diminishing strength in the Party apparatus had spread to the scholarly community, it might have been taken as a sign that the balance of power at the top was changing. Still, the meeting was organized as a session of the Agricultural Academy and not as a full-blown meeting under Party sponsorship.

Though over seven hundred people attended, the names and positions of the participants in the session did not indicate that a major policy decision was under way. Among philosophers, for example, both Beletskii and Mitin spoke in defense of Lysenko and against the "idealist" tendencies of the "Mendelian-Morganists." But Kedrov, the editor of the major philosophical journal, and Aleksandrov, the director of the Institute of Philosophy, both of whom would have been more reluctant to criticize the geneticists, did not even attend the meeting. Whereas academicians in a full range of fields had attended the philosophy discussion, the biology session was limited to those with a special interest in biology, agriculture, or the philosophy of science.[63] In every sense, then, the biology session was trumpeted as a scientific meeting. The role of the Party was hidden. But this did not improve the tone of the discussion. At the philosophy meeting, disagreements had cropped up without necessarily causing accusations of unpatriotic, anti-Soviet, or pseudoscientific behavior. At the Agricultural Academy session, it seems as though even the slightest disagreement with Lysenko on any point spawned vicious, personal, and dangerous rebuttals. Even in comparison with Andrei Zhdanov—whose name is rightly associated with the aggressive and blunt campaign to assert Party discipline in literature and the arts—Lysenko came across as an uncompromising zealot. The philosophy discussion had sought to forge a Party line; Lysenko was out to dictate a scientific theory.

Lysenko was not alone. He managed to fill the hall of the House of Scientists with eager followers, including the newly appointed members of the Agricultural Academy. Unlike the philosophy discussion, where Zhdanov's speech had come a number of days after the meeting had begun, Lysenko's speech opened the Agricultural Academy meeting on the evening of July 31. P. P. Lobanov, the deputy minister of agriculture and one of the recent appointees to the Agricultural Academy, presided. The next day participants in the meeting went to Lenin Hills outside Moscow to witness some of Lysenko's agricultural accomplishments. On August 2 the meeting resumed, and speeches were delivered in response to Lysenko's lecture. The structure thus resembled a large Soviet-style scientific meeting.

The content, of course, was a different matter. When Stalin removed references to "bourgeois" and "proletarian" science from Lysenko's speech, he also eliminated a means of equating Michurinism with the official ideology. Without references to class, how were Lysenko and his allies supposed to distinguish their work from that of the geneticists? They relied on other familiar approaches. They insisted, for example, that their work was utilitarian and oriented toward practice, while ridiculing the abstract values of their opponents. They also continued to contrast "foreign" science and "Soviet," patriotic science. These distinctions pointed to the conclusion that was of fundamental importance to Lysenko and his followers: their views were scientific, and those of their opponents were pseudoscientific.

Since the late 1920s the Party had been insisting that Soviet science serve the state. In simplest terms this meant that theoretical work was acceptable only as a step toward applying those ideas to some practical problem. Science for science's sake was not good enough; all science had to play a role in socialist construction. Although theoretical work continued, Soviet science was not primarily about international prestige, theoretical elegance, or even the attainment of "truth." Instead of these standards, science in the Soviet Union was to be judged by its practical results.

In the case of biology, one of the clearest avenues for application was agriculture. The frequent famines in Russian and Soviet history only helped to clarify the importance of this connection. Any work in biology with potential relevance to agriculture—whether genetics or soil science—was to be judged by its contribution to feeding Soviet citizens. Party propaganda also emphasized that the Soviet Union was the best place for such applied work to take place. After all, the USSR claimed to have created the ultimate conditions for science to prosper, freeing it from the bourgeois and capitalist influences that would distract scientists and science from the goal of transforming nature to meet the needs of the people. That the conditions for

Figure 5. Trofim Lysenko at the lectern during the Argricultural Academy Session of 1948. Courtesy of RGAKFD.

science in the Soviet Union were better than anywhere else in the world was a given of Stalinist politics.

Though it seems counterintuitive at first, the emphasis on practical results gave Lysenko an advantage. Whether he had actually improved grain harvests or helped to create new varieties of plants was secondary to the fact that his professed goal was to use his science to support directly Soviet agriculture and the building of socialism. At this early stage, geneticists suffered from something that in the Soviet system was worse than being "wrong"; Lysenko attacked them for emphasizing theory at the expense of practical results. They asked the state to spend resources on laboratories organized around problems that had no clear application to socialist construction. This meant that "Mendelian-Morganism" was inherently "antiscientific."

The emphasis on practical results as a measure of who was correct in science arose consistently during the meeting. People spoke of the "miraculous transformations" of plants and contrasted this with the "sterile and effete" efforts of the Mendelian-Morganists.[64] Even when Lysenkoists admitted that mutations had been obtained with fruit flies, they claimed that the experiments with *Drosophila* were useless by their very nature. How would fruit flies help Soviet agriculture? As the logic went, only production could justify science. One speaker put it succinctly: "Our Soviet science cannot fence itself off from production. It must be connected with production by constant living ties and

serve the cause of strengthening the economic and political might of our country. Only such a science justifies its existence."[65]

Practical science was equated with the ability to transform nature. One of
the speakers, a director of a cattle-breeding station, was fearful of the results
of Mendelism-Morganism, which he said was "reactionary to the core since
it is bound to lower the role of Soviet man; this theory wants to make us bend
our knees to nature; it tries to convert Soviet man into a passive appendage
of nature, a placid contemplator of nature who humbly waits for gifts and
favors from her." In contrast, Michurinism was the "great transformer of
nature."[66]

Mendelism-Morganism was also considered pseudoscientific because it emanated from abroad. As the historian Nikolai Krementsov has clearly shown,
the geneticists committed a grave tactical error. In the immediate postwar
period they had cited their prestige in the "international scientific community" in defense of their science. By 1948, however, being associated with
foreign science was clearly a sign of weakness and error. If the Soviet Union
had created the best conditions for science, it only followed that international
science was likely tainted by "bourgeois" and "capitalist" assumptions.[67]

Lysenko, in contrast, looked almost exclusively to Russian and Soviet scientists when tracing the genealogy of his ideas. Whereas the geneticists were
deemed guilty by association with Western scientists who held views similar
to theirs, Lysenko could revel in the fact that no one outside the Soviet Union
seemed to take his work seriously. This was proof to him that his approach
was correct. Westerners could not understand the scientific advances under
way in socialist Russia. In this manner, Lysenko distanced himself from the
nation's ideological enemies.

The speakers at the meeting followed Lysenko's lead. One of the recent
appointees to the Agricultural Academy enthused: "We may justly be proud
that distinguished biologists lived and worked in pre-Revolutionary Russia
and in the Soviet Union. . . . In the eighteenth and nineteenth centuries
there were no Weismanns, Batesons, Lotsys or Morgans in our country." Another speaker identified Mendelism-Morganism as reactionary in part because
it "sounded strange in Russian." The philosopher Mitin encouraged the audience to feel proud that biology had been pushed forward by "Michurin, a
Russian, Soviet scientist," as opposed to the "cosmopolitans."[68] Prezent saw
the roots of Soviet biology in the pre-Revolutionary philosophies of Herzen,
Belinsky, Chernyshevsky, and others, while a spokesman from the Ministry of
Education emphasized that Soviet people had a special place in their heart
for "distinguished Russian scientists."[69]

In contrast, the Soviet followers of Mendelism-Morganism were seen as
having allowed "alien trends from abroad" to "penetrate into our science."
One speaker even feared that Western theories "lead to a loss of faith in the
possibility of communism being victorious in our country," while in contrast "a

crimson thread" ran through the work of Michurin and Lysenko.[70] Apparently without any contradictions, Michurinism had managed to gather the best of both the pre-Revolutionary tradition and the Soviet worldview.

The Lysenkoists singled out Dubinin as particularly unpatriotic because of his article in *Science* and because he had worked on fruit flies during the war. They depicted him as "divorced from life" and oblivious to the needs of the country. They accused Boris Zavadovskii of endorsing an "international language for science." As Mitin presented it, Soviet biologists had two choices. Either they could "further develop our Soviet, consistently materialist, Michurin trend in science . . . or slavishly accept the unscientific, idealistic concepts of foreign bourgeois 'authorities' who are hacking at the roots of the theory of evolution."[71]

According to Lysenko's followers, the geneticists not only kowtowed to the West; within Soviet society itself they occupied a stratum that hinted at bourgeois values. They were "private gentlemen," conducting their work in laboratories far away from the working people. Not only did they ask "theoretical and abstract" questions without regard for practical concerns, their means of answering those questions were helplessly "academic." Lysenko, on the other hand, emphasized that the ultimate test of scientific truth was its acceptance by the masses, by the "army of people on the farms" and the "practical folks" who worked the soil. This was a standard that made many of the geneticists uncomfortable. What did a farmer know about science? But the geneticists' methodological concerns were used against them to reveal class bias. Time and again, Lysenko's supporters cited the working farmers as judges of what worked and what did not work in a practical sense.

A large number of the speakers at the conference were practical workers who sang the praises of Lysenko's methods. The geneticists did not galvanize such "men of the people" to defend their views. One speaker complained that "exalted science" suppressed "modest practical workers." Whereas Lysenko was "an energetic organizer of the masses, supported by the work of millions of collective farmers," his opponents were "scholastics" and "arm chair philosophers," out of touch with the people on the farm. Every collective farmer was praised as an experimenter, and it was the farmers' support that proved that Lysenko's science was correct. It was wrong to attempt to "divide science from callused hands." One speaker banged the point home: "Just try, if you can, to swing this tractor driver round to the position of bourgeois agrochemistry! No, he has already put his weighty tractor driver's hand into the scales of the old scientific controversy . . . and his word is decisive."[72] This was offered as a stronger argument than anything geneticists could come up with in the lab.

Another means of showing that the geneticists were pseudoscientific according to Soviet standards was to show the "unscientific" tactics the geneticists used in defense of their work. The Lysenkoists accused them of monopolizing biological research establishments, scientific publications, and biology teaching, as well as refusing to engage in discussion, preventing the exploration of Michurin's theory, and arrogantly dismissing views that were not compatible with their own. Ironically, the Lysenkoists leveled these accusations using similar tactics. When they recognized that perhaps they too were now intimidating their scientific opponents, they absolved themselves of wrongdoing by claiming that they did it with scientific truth on their side.[73]

The Lysenkoists' conclusions were clear. Scientific truth came from practical results; it was recognized and supported by everyday Soviet folk; it was distinct from truth in the West; and it was not defended with extrascientific practices. These, then were the central arguments presented by Lysenko's supporters in the August session.

Of the fifty-six speakers at the conference, only nine dared challenge Lysenko's tone and thesis in the slightest way. These nine scholars had numerous disagreements among themselves, further diminishing their strength and adding to the sense of victory for Lysenko and his followers. Although the session began as a limited forum for discussing Lysenko's speech, by the end it was receiving national attention. On August 4, four days after the start of the meeting, *Pravda* began publishing portions of the speeches and continued to do so for a week. What had apparently begun as a scientific session had quickly evolved into a Party-organized rout of genetics. There should be no doubt that Stalin intended the session to conclude that the "future [of biology] belonged to Michurin." But he did not expect to show his own and the Party's roles in this decision. Scientific truth arrived at by scientists working within a worldview defined by the Party was different from the Party dictating the truth to scientists. Over the course of the discussion, however, the rebuttals of the few dissenters forced the focus of the session to change. By the end, Lysenko's authority rested not on claims of practicality, patriotism, or theoretical rigor, but on the fact that the Central Committee (meaning Stalin) had read and approved his speech.

Dissenters responded to the declarations of the Lysenkoists in a variety of ways. They tried to show that their work was practical; they emphasized the role of Russian science in their work and accepted that the "future belongs to Michurin" while trying to reclaim his legacy from Lysenko. They also tried to separate political naïveté from scientific error by apologizing for "unpatriotic" behavior. Finally, they pointed to the unscientific tactics used by Lysenko and his cohort over the years. In the end, the most important form of opposition to Lysenko was to refuse to accept, until more substantial evidence was presented, that the Party and Stalin supported Lysenko's position.

At the end of the first session on August 2 Lobanov, who was presiding, implored the "formal geneticists" to take the floor. Perhaps he assumed that the geneticists would remain silent, much as the Lysenkoists had held their tongues at the MGU meeting of 1947. But Iosif Rapoport, a researcher at the

Institute of Cytology, Histology and Embryology of the Academy of Sciences, was anxious to rebut Lysenko's speech. Rapoport had fought on the front during the Second World War, had been involved in the liberation of the Crimea, and, along with other injuries, had lost an eye in battle. He did not cower before the Lysenkoist majority surrounding him.[74]

Rapoport agreed that Soviet science was superior to and more practical than science in capitalist countries, but he warned against being overly confident. Insisting that truth emerged from scientific conflict and that this conflict must be allowed to take place, Rapoport launched into a criticism of Lysenko's speech. He defended genetics by noting that the modern founder of the theory of the gene was Darwin. He also cited Lenin to bolster his argument that science from the West should be critically and creatively assimilated, not dismissed. Genetics, Rapoport insisted further, had many practical applications. He also added that Lysenko's adaptation of Lamarckism led to errors. Needless to say, Rapoport became the object of constant ridicule for the rest of the meeting.[75] Like-minded scientists were allowed to take the floor only after two more days of pro-Lysenko speeches.

Some dissenters tried to outline their differences with Lysenko on certain points while agreeing with his general thesis. Il'ia Poliakov, for instance, decried the anti-Darwinian, bourgeois biology of the West. "Biology," he asserted, "is a reflection of the acute, intense class struggle." But he could not accept Lysenko's emphasis on the environment as the sole factor in evolutionary change: "If we take it that the environment alone causes adaptive change in the organism, it will inevitably lead us to theology." When he questioned Lysenko's views on intraspecies competition, Lysenko interrupted him. Prezent joined in as well, and Poliakov became flustered.[76] Alikhanian also tried to argue that the environment could influence the nature of the gene but that genes, nonetheless, existed. Lysenko challenged him too, and though he ended with a patriotic speech that earned him applause, he failed to show up after a break to answer questions.[77]

Zhebrak also spoke, but concentrated almost exclusively on his technical data. His response to the question of what experimental genetics could contribute to industry indicates that he failed to grasp the implications of Lysenko's speech: "In my opinion these investigations raise the level of our Soviet science. They are in unison with Comrade Stalin's instruction that the task of Soviet scientists is not only to catch up with the achievements of science of other countries, but to outstrip them." Zhebrak seemed to call for improving science for science's sake, and this was not acceptable. Practical applications were the primary measure of scientific merit.[78]

The most prominent scientists among the dissenters to address the meeting were Academicians P. M. Zhukovskii, I. I. Shmal'gauzen, V. S. Nemchinov, and B. M. Zavadovskii. Zhukovskii, a member of the Agricultural Academy, refused to be interrupted by Lysenko and even went so far as to declare that "one should worship" at Mendel's grave. Zhukovskii could not understand why there was a "veritable craze" at the meeting to defame Mendel, when all he had done was make some sound observations about the laws of heredity in peas. Mendel had never written about evolution, nor had he excluded external factors from influencing heredity. There was no need to assume that Michurinism and Mendelism were in contradiction. Zhukovskii called for unity among Soviet biologists, noting, "We are all Soviet citizens, and we are all patriots. Some of us went personally and others sent their sons to the front. We all fought for our country, and should we really allow things to reach such a point that people refuse to greet Professor Zhukovskii when they meet him?" His speech was met with applause.[79]

Shmal'gauzen defended himself as a materialist critic of Weismann and other geneticists. As he presented it, his disagreement with Lysenko was similar to that of Alikhanian and Poliakov; he believed that the external environment played a role in evolution but insisted that the internal organism played a bigger role. Throughout the meeting his book *Factors of Evolution* had been criticized for not mentioning Michurin or other Russian plant breeders. Shmal'gauzen argued that the book simply had nothing to do with the question of breeding and that elsewhere he had cited Michurin. Like Zhukovskii, Shmal'gauzen did not directly attack Lysenko and instead sought some form of reconciliation.[80]

The meeting was most forcefully disrupted by the speeches of Nemchinov and Zavadovskii. In fact, recently declassified archival material reveals that their speeches challenged Lysenko much more than is evident from the published record. When the official report of the speeches was published, these two were significantly abbreviated, with some sections removed entirely. The published versions of the other speeches closely resemble the stenographic report found in the archives and usually have only minor stylistic corrections. The differences between what was actually said and what was subsequently reported provide a good indication that Nemchinov and Zavadovskii did not speak according to the rules of how the meeting was supposed to progress.

At the time of the meeting Nemchinov, who was an economist and statistician by training, directed the Timiriazev Agricultural Academy, where Zhebrak conducted his research on genetics. He was also a member of both the Academy of Sciences and the Agricultural Academy. Despite his prestige, a number of the Lysenkoists had accused him of purging Michurinists from the ranks of the Timiriazev Academy. Nemchinov had not originally planned on participating in the meeting, but the repeated attacks against him and his academy changed his mind. When badgered about his support for Zhebrak and his work, Nemchinov said that he had been consistently critical of Zhebrak's unpatriotic behavior, such as his articles in American journals, but supported his scientific opinions. There was a difference between political mistakes and scientific mistakes. Someone from the audience then pressed him on the core issue of the meeting: "Is the chromosome theory one of the great discoveries of science?" Nemchinov responded: "Yes, and I can repeat it: by understanding the effect of tampering with and manipulating chromosomes, we can transform nature and by using the change in the generations of plants and animal organisms we can alter their form. I think that the chromosome theory of heredity is one of the great discoveries in biology." According to the archival record, this statement was greeted with applause. In the published version, Nemchinov's response is abbreviated and the applause is renamed "commotion."[81]

When the audience further challenged Nemchinov about his authority to reach such conclusions, he declared that though he was not a biologist, training in statistics led him to believe in the validity of the theory. This time his statement met with "applause and commotion." According to the unpublished record, Nemchinov went on to defend the existence of chromosomes as well as Zhebrak's right to work at the academy. He repeated that Michurinists should continue to work there as well. Nemchinov tried to resist Lysenko's campaign to outlaw work in genetics completely and to label "anti-Soviet" any idea that did not correspond to Lysenkoist theories. The fact that at least part of the audience applauded Nemchinov's effort might explain why the published version was so thoroughly edited. When someone in the audience protested that the meeting had deteriorated into an interrogation of Nemchinov, Nemchinov agreed that things had indeed come to that.[82]

The greatest challenge to Lysenko's control of the meeting concerned his relationship to the Party, not whether his ideas were correct or not. B. M. Zavadovskii, who had supported Lysenko in the 1930s, questioned the conditions under which the discussion was being held and the position of the Party on the issues being debated. If the Party did not support Lysenko, then Zavadovskii concluded he could openly engage Lysenko in scientific argument without fear of an official reprimand. Zavadovskii's speech was also severely edited before publication. The archives now show that he was in fact much more adamant about pressuring Lysenko on the Party's role than was previously known:

Having already arrived in Moscow and received my invitation to the session, I turned to the Central Committee of the Party with a question: how should all this be understood? I was prepared to speak, but received an explanation that though the Central Committee was not against my speaking, it also did not require me to speak. I thus understood that the conference was obviously taking place without the agreement or at the very least without the participation of the Science Section of the Central Committee of the Party. Inasmuch as it was up to me to decide, I believed that it would be more correct and rational to ask the leaders of our Party to organize, with the participation of the Secretariat of the Central Committee, a special discussion on the state and needs of biological and agricultural science. This discussion would be similar to the ones on philosophy and literature and the arts. This would give the best opportunity for free and thoughtful statements for those who participate in the building of Soviet science so that there would not be the atmosphere of premature conclusions which has at different times appeared on the pages of *Socialist Farming* and the *Literary Gazette*. In this way we could avoid the prior indiscriminate branding of everyone who in some way disagrees with T. D. Lysenko as a Weismannist, Malthusian, and perpetrator of other deadly sins.[83]

Zavadovskii emphasized that the Party was not taking an active role in the discussion. He justified his own participation by noting that after reading Lysenko's speech in *Pravda*, he believed it was his duty as a Party member to take on the battle even under very unfavorable conditions. He proceeded to declare himself a loyal Michurinist and to accuse Mitin and Lysenko of diverging from Darwinism and from Michurinism. He insisted that it was possible to disagree with Lysenko and not be a "formal geneticist." Like Zhukovskii, he defended Mendel and other trends in science that he believed needed to coexist with Lysenko's views. His conclusion reiterated his belief that he was within his rights to challenge Lysenko so long as the Party had not informed him otherwise: "For the time being, however, I am justified in thinking that to the best of my strength and ability I am defending the Party's position as to the struggle for its general line in solving the problems of Darwinian theory."[84] The tone of the dissenting opinions, especially as they come across in the stenographic record in the archive (as opposed to the published report), makes it clear that the discussion alone did not succeed in establishing Lysenko's hold over biology. The lack of a strong Party presence confused participants who assumed that an official discussion would resemble the one held in philosophy the summer before. It was now up to the Party to inform the public what had already been decided behind the scenes.

Zavadovskii's speech clearly caught the attention of the Central Committee. Agitprop chief Shepilov immediately wanted to know what conversation had taken place between Yuri Zhdanov and Zavadovskii. Zhdanov's assistant reported to Shepilov the following information:

At 11:00 p.m. on July 31 I was present when Zavadovskii called Y. A. Zhdanov and asked to meet with him immediately. Comrade Zhdanov answered that he was unable to meet with Zavadovskii. Zavadovskii asked if he could speak at the session about Academic Lysenko's talk. Comrade Zhdanov answered that speaking at the session was a matter for Zavadovskii himself, and that he should decide for himself whether he should speak. Comrade Zhdanov said, "I can neither object to speeches at the discussion, nor recommend them." On that note the conversation ended.[85]

Within hours of Zavadovskii's speech Shepilov sent Malenkov the excerpt where Zavadovskii mentioned that the Agricultural Academy meeting was taking place without the participation of the Central Committee of the Party.

Shepilov interpreted Zavadovskii's words as provocative and suggested one of three actions be taken. First, Shepilov offered to give a short speech at the session to counteract Zavadovskii's. Second, he suggested that he give a longer talk, not just addressing Zavadovskii's speech, but providing a more general statement in line with Lysenko's views and critical of the fundamental views of Weismannism. He believed he could prepare such a speech in one day, since he had studied the problems in genetics previously and had examined them intensely over the past two months. Finally, Shepilov recommended that if it was decided that he should not participate, at least Lobanov could criticize Zavadovskii.[86] Obviously, a speech from the head of Agitprop would have severely altered the style of the meeting. Whether it took the form of a short or a long speech, any direct participation by Shepilov would have made blatantly clear that Lysenko's authority derived from the Party, not from scientists. Malenkov rejected Shepilov's offers and probably consulted with Stalin on what was to be done.

Their solution became clear on the morning of August 7. First, *Pravda* printed Yuri Zhdanov's repentant letter to Stalin in which he recognized the errors in his anti-Lysenkoist talk in April.[87] This clarified that a key element of support for genetics in the Party no longer existed. The Party went further than repudiating Zhdanov's stand, adding a crucial sentence to Lysenko's closing remarks at the conference. Before beginning his prepared statement, Lysenko read a note written hastily in his own hand and edited by Stalin: "The question is asked in one of the notes handed to me: What is the attitude of the Central Committee of the Party to my report? I answer: the Central Committee of the Party examined my report and approved it."[88] The statement was greeted by "stormy applause" and a standing ovation. It was now clear to one and all that the Party supported Lysenko.

The ultimate arbiter of scientific truth was thus the Party itself. More than the voice of the people or the importance of practical applications, the Party determined what was correct. But this was a role that the Party asserted in surprisingly obtuse ways. Despite the dominance of Lysenko's supporters at the conference, the rhetorical advantages listed above, the organizational control of the event, and the attention the conference received in the press, the discussion in biology was only really concluded on the last day of the conference. The publication of Yuri Zhdanov's letter and Lysenko's reference to the role of the Central Committee now clarified the Party's stand. Further dissent against Lysenko would be considered an affront to the Party as well.

Immediately following Lysenko's closing speech, Zhukovskii took the floor to declare his allegiance to Lysenko and Michurin and pledged to devote his energy to fighting "foreign reactionaries." Since the Central Committee of the Party had "drawn the dividing line between the two trends in biological science," Zhukovskii recognized that his previous speech had been "unworthy of a member of the Communist Party and of a Soviet scientist." He emphasized that he was joining the ranks of Lysenko's supporters "as a sincere member of our Party."[89]

Alikhanian quickly followed suit, stating, "I, as a Communist, cannot and must not, in the ardor of controversy, obstinately oppose my personal views and concepts to the onward march of biological science." He added that he would adhere to this new belief as a teacher and researcher: "From tomorrow on I shall not only myself, in all my scientific activity, try to emancipate myself from the old reactionary Weismann-Morganist views, I shall also try to reform and convince all my pupils and comrades." He echoed the Lysenkoists' patriotism as well, declaring: "And only in our country, the country with the most advanced and progressive world outlook, can the seedlings of the new scientific trend develop."[90]

Finally, Poliakov endorsed Lysenko's scientific leadership and accepted that "Michurinist theory defines the basic line of development of Soviet biological science." He vowed to "assist the Party in exposing the reactionary pseudo-scientific rot which is disseminated by our enemies abroad" and to devote his strength "to the promotion of the great Michurinist theory."[91] The duplicity of Zhukovskii, Alikhanian, and Poliakov's statements is obvious but understandable given the context of Stalinist Russia. Others, including Nemchinov, Zavadovskii, and Rapoport, did not recant. Still, what had been a scientific discussion had been transformed into a question of loyalty to the Communist Party.

The Central Committee did not take long to implement changes on Lysenko's behalf.[92] In a series of meetings during the week of August 9, the Orgburo and the Secretariat approved measures to replace "reactionary supporters of Mendelian-Morganism" with Lysenko's loyal cohort in research institutions, teaching establishments, editorial boards of scientific journals, and elsewhere. They also organized a campaign to promote "Michurinist biology" in the press. On August 9, for instance, the Orgburo passed resolutions to dismiss Nemchinov as director of the Timiriazev Academy, Shmal'gauzen as the chair of Darwinism at MGU, Iudintsev as the dean of biology at MGU, Zhebrak as chair of genetics at the Timiriazev Academy, M. E. Lobashev as dean of biology at Leningrad State University, and Polianskii as that university's prorector. They were replaced by Lysenko and his followers.[93]

On August 11 the campaign was well under way as the Secretariat continued to dismiss people and set about drafting plans for personnel changes in the Ministries of Education and Agriculture, the Biology Division of the Academy of Sciences, and the Stalin Prize Committee. It also restructured publishing plans at the state publishing house and teaching plans at the universities and in schools.[94] On August 12, *Pravda* printed a front-page editorial celebrating the victory of Michurinist biology over the "reactionary bourgeois biology" of the West, while praising the work of Michurin and Lysenko as materialist and cutting edge.[95] By mid-August Sergei Kaftanov, the minister of education, reported that twenty-five "formal geneticist" deans from institutions around the country had been replaced by "Michurinists."[96] Throughout the fall of 1948 institutes, universities, and academies held meetings displaying their vigilance in upholding the Central Committee's decrees.[97]

In many ways, given the political context of the Soviet Union during the late 1940s, Lysenko's victory was overdetermined. He was a muzhik; his opponents, inasmuch as they could be said to be a single group, had traces of "bourgeois" values. His science promised immediate improvements in agricultural production; his opponents were hard pressed to show the relevance of their work for the economy. Lysenko had a strong antipathy for the West; his opponents relied on Western science as their standard. Lysenko sought the roots of his science in Russia; his opponents were following a path established by foreigners. Lysenko viewed his work as steeped in dialectical materialism; his opponents strained to connect their work with the official philosophy. Lysenko's political clout emanated from Stalin and Malenkov; his opponents worked through Zhdanov, whose political power was ebbing. It is difficult to pinpoint any single factor as decisive in Lysenko's ability to gain the Party's trust. Lysenko was able to make his work appealing on all fronts.

Stalin's editorial comments and written exchanges with Lysenko suggest that one of the most important reasons the leader threw the full weight of the Party behind "Michurinist biology" had to do with ideas, not politics. As odd as it seems to those of us who now accept the practical rewards of modern genetics, Stalin apparently believed that Lysenko was fundamentally correct about the inheritance of acquired characteristics and man's ability to use that knowledge to control nature.[98] It is a sign of remarkable intellectual arrogance that Stalin would take it upon himself to tell scientists which ideas in their field were correct and which were not. Just the same, he also recognized that scientific theories had to at least appear to emerge victorious from a process that included the exchange of opinions, or discussions. This is the most logical way to explain his reluctance to simply pass a decree supporting Lysenko. "Correct," "objective" ideas could receive the Party's support, but they could not appear to have been dictated.

As Stalin tried to apply this "victory" to other fields of science, he found himself in the fall of 1948 in a strange position. Andrei Zhdanov, the lieutenant on whom he had relied for intellectual comradeship, died of heart failure on August 31. In his place stood Malenkov and Beria, neither of whom had given much time to theoretical issues. It seemed only natural that they would latch on to the xenophobia of Zhdanov's campaign rather than the challenge of reinvigorating Marxist-Leninist doctrine. Among the leadership, Stalin alone seemed concerned about the theoretical problem of keeping the official ideology up to date with the latest advancements in science. He could also rely on one key figure within Agitprop—Yuri Zhdanov—who kept his position as the head of the Science Section despite his father's death and Lysenko's victory. His training in science and philosophy would make him a useful adviser as ideological disputes spread to other disciplines. Stalin's respect for the young apparatchik even took on a personal angle: he encouraged his only daughter, Svetlana, to marry Yuri Zhdanov, and in the spring of 1949 the young couple obliged.

The obvious tragedy of Lysenko's rise and the demise of genetics has made it difficult to recognize that this episode, ironically, signaled a tremendous boost for the prestige of "science" in the Soviet Union. For those unaware of the erroneous assumptions behind Lysenko's views—that is, for the majority of the readers of *Pravda* and for almost everyone in the government and Party apparatus—the Lysenko affair was a celebration of science and the ability of science to flourish in the USSR. Lysenko wisely shrouded his claims within the rhetoric of class struggle (his biology was proletarian), the growing national struggle (it was homegrown), and the ideological struggle (it was dialectical materialist.) But Stalin touted his ideas as scientific and gave him authority in the Soviet Union as a scientist, not as a Party functionary. In the wake of such a dramatic declaration of harmony between the Party and science, one question remained for Stalin, his assistants in the Central Committee, the scientific administrators in the Academy of Sciences and the Ministry of Education, and the Soviet scientific community: how could the lessons of "the situation in biology" be applied to other fields of scholarship?

CHAPTER 4

"WE CAN ALWAYS SHOOT THEM LATER"

Physics, Politics, and the Atomic Bomb

In early August 1948, across Moscow from where the Agricultural Academy session was taking place, a plan was germinating to bring together the USSR's physicists. On August 3 B. M. Vul, the head of the Physics-Mathematics Division of the Academy of Sciences, proposed gathering academy and university physicists for a pair of meetings during the winter break in the school year. At first, the purpose of the meetings had little in common with either the philosophy discussion or the Agricultural Academy session: the goal was to increase scholarly exchange between the academy physics institutes and the universities' physics departments. The modesty of the proposed meetings reflected the limited nature of the problems to be addressed: the agenda included pedagogy, laboratory techniques, and the design of physics courses for engineers. The fact that in August 1948 neither the president of the Academy of Sciences nor the minister of education became directly involved further indicates that there was little intention of making the event a matter of national importance.[1]

In the aftermath of the Agricultural Academy session, however, the fairly innocuous plan to bring together physicists from the academy and the universities quickly evolved into a forum for addressing ideological problems in Soviet physics. In the months following Vul's original proposal, the two meetings merged into a plan to hold a single All-Union Conference of Physicists, which many participants hoped would expand the "achievements" of the philosophy discussion and the Agricultural Academy session. An initial effort to reconcile differences between the Academy of Sciences and the universities and to coordinate research and teaching developed into a broad discussion of physics, philosophy, and Soviet patriotism.

The stakes in physics were high. In the aftermath of the U.S. atomic bombing of Hiroshima and Nagasaki in August 1945, Joseph Stalin made the development of a Soviet atomic bomb the USSR's top priority. Though Stalin placed Lavrenty Beria, the head of state security, in charge of administering the colossal undertaking, he also understood that physicists would play an essential role in breaking the U.S. atomic monopoly. That fall, the physicist Peter Kapitsa wrote to Stalin to complain about the way Beria was running the Soviet atomic bomb project. As one of the most prominent scientists in the USSR, a member of the Academy of Sciences, a Hero of Socialist Labor, and a member of the Special Committee charged with developing nuclear weapons, Kapitsa confidently explained to Stalin that scientists would not work well unless political administrators treated them with more reverence. He was especially concerned about Beria's arrogant treatment of scientists working under him. "It is true, he [Beria] has the conductor's baton in his hands," Kapitsa wrote to Stalin. "That's fine, but all the same a scientist should play first violin. For the violin sets the tone for the whole orchestra. Comrade Beria's basic weakness is that the conductor ought not only to wave the baton, but also to understand the score. In this respect Beria is weak." Soon Kapitsa left the Special Committee, but, as the historian David Holloway has pointed out, in some ways his recommendations were heeded. As the scientific director of the atomic project, another physicist, Igor Kurchatov, served as scientific "concertmaster" to Beria and other administrators and in turn earned their respect.[2]

Kurchatov's access to powerful political leaders tended to help Academy of Sciences physicists more than others. Like Kapitsa, Kurchatov had begun his training under Academician Abram Ioffe in Petrograd and had risen through the ranks primarily as an academic researcher, rather than as a university professor. Before the Second World War, many prominent Soviet physicists taught at universities while conducting research at the Academy of Sciences.[3] But during the war the university physics colleges (fakul'teti) and the physics institutes of the Academy of Sciences were moved to different locations from each other, so that scientists with appointments as both researchers and teachers had a decision to make. Most chose to stay with the academy and continue their war-related research. When the war ended, many did not return to the universities. This was especially true in Moscow, the home of several important academy physics institutes. One of the reasons physicists chose to stay away from Moscow State University was that its Physics College was increasingly coming under the control of Party members who, among other things, were appointing department chairs for political rather than scientific reasons.[4] During the war, for instance, Igor' Tamm, a researcher at the Physics

Institute of the Academy of Sciences and a future Nobel laureate, was not chosen as the head of the Theoretical Physics Department, losing out to a relatively obscure junior member of the faculty. Some of the most able physicists left the university in the middle 1940s, including Tamm, S. E. Khaikin, a premier lecturer in mechanics, and Academician M. A. Leontovich, an expert in plasma physics. After the death in 1944 of another gifted teacher, L. I. Mandel'shtam, many of his best students also left the university for positions in the academy.[5]

With academy physicists dominating the atomic project and benefiting from the resources being spent on it, the split worsened. Kurchatov plucked the best graduate students from the university to work for him. The highest echelons of the Party apparatus favored the academy physicists as well. When one MGU professor, N. I. Kobozev, complained to Malenkov that his chances of winning a Stalin Prize had been sabotaged by an "unproductive" and "unprincipled" academician, Central Committee workers concluded that Kobozev had no basis for his denunciations.[6] Academicians with connections to the atomic project had more influence on professional decisions than MGU professors. In 1946 a group of physicists working at academy institutes received positions either as corresponding or full members of the academy, while those from MGU were passed over. In a top secret memo to the Politburo commission overseeing the elections to the academy, Kurchatov and key administrators working on the atomic project endorsed the candidacies of academy scientists working on the bomb.[7] Similarly, physics journals seemed to be academy enterprises, with academy physicists dominating editorial boards, writing most of the articles, and referencing and reviewing each other's work. Physicists outside the academy felt denigrated or ignored.

Academy dominance and the influence of the atomic project were not the only reason MGU physicists were in a subordinate position. Without physicists such as Tamm, Fock, and Mandel'shtam on the faculty, the Physics College at MGU was dominated by less prominent but more politically engaged scientists such as V. F. Nozdrev, the secretary of the college's Party Committee. A repressive atmosphere at MGU that discouraged discussion of the most recent work of Western scientists added to the void left behind by the exodus of academicians.[8] A Central Committee report in 1945 suggested that the Party organization at MGU was driving physicists away:

> Over the last few years, young scholars, students and graduate students at MGU, have shown conceit, arrogance, and disdain toward such important scientific centers as the Academy of Sciences, toward the activities of older scientists, and toward the accomplishments of foreign science. The Party leadership of MGU has an overly exaggerated sense of its own accomplishments, which more than once has led a group of university professors, who aspire to a leadership role in physics and chemistry, to launch baseless attacks against the Academy of Sciences and various prominent scientists.[9]

In short, the institutional chasm between the academy and MGU provided a base for disputes among physicists.

The Central Committee attempted to reduce the growing tensions by ordering the university to invite specialists from the academy to teach courses. This caused even more problems. Party leaders at MGU welcomed two physicists who had little time to teach: Kurchatov and Sergei Vavilov, the president of the Academy of Sciences. But the university physicists remained adamantly opposed to the influence of Kapitsa, Fock, Ioffe, and other academicians in the university's Physics College. The Party Committee secretary Nozdrev wrote to Malenkov and Moscow Party chief G. M. Popov about the political danger Kapitsa, Fock, and others presented to the students of the university. Nozdrev accused Kapitsa of propagating the idea of the "dictatorship of the intelligentsia" and calling for the subordination of MGU and Leningrad State University to the Academy of Sciences. Nozdrev further asserted that Kurchatov's mentor Ioffe and others were teaching idealism and creating anti-Soviet sentiment among the student body. The minister of education, the chemist Sergei Kaftanov, responded with a long letter to Malenkov defending Kapitsa and noting that the head of the Science Section of the Central Committee, Sergei Suvorov, as well as prominent administrators in the atomic project could attest to the numerous mistakes in Nozdrev's letter.[10] Again, powerful patrons in the Party and government sided with academy physicists.

The situation deteriorated to the point that university physics and academy physics seemed to be in constant conflict. In a letter to Stalin written in October 1947, the former dean of physics at MGU explained, "A distinctive 'ideology' is developing and is spreading a particular theory of university science in opposition to 'unhealthy' academy science."[11] As the events of the next few years would show, university physicists took advantage of the broader ideological campaign to question the patriotism of academy physicists and to point out the so-called bourgeois and idealist elements in their work. The academy dominated the atomic project, but it also employed many physicists who had worked and studied abroad in the 1920s and 1930s (making them susceptible to charges of subservience to the West) and who openly endorsed Western physical theories. Which side would Stalin and the Party take in the disputes between the academy and university physicists? The discussions in philosophy and biology suggested that the university physicists had some clear advantages.

When Andrei Zhdanov presided over the discussion of Aleksandrov's *History of Western European Philosophy*, in 1947, those in attendance understood that philosophy, as defined by the Party, was to permeate all intellectual spheres. Coming on the heels of his attacks in literature and the arts and a revived campaign against foreign influence in Soviet society, Zhdanov's statements made clear that the effort to strengthen postwar Soviet ideology had spread. Some scientists, most notably Lysenko, insisted that the new ortho-

Figure 6. Igor Kurchatov, scientific director of the atomic bomb project. Courtesy of the Kurchatov Museum.

doxy was applicable to the natural sciences. So physicists of all stripes listened attentively to Zhdanov's speech, searching for clues that might indicate the Party's intentions for their field.

Given Zhdanov's position within the Party, it is not surprising that his speech met with universal praise. But Suvorov, the head of the Central Committee's Science Section and Zhdanov's direct subordinate, was troubled by the paragraph of his speech in which he declared, "The Kantian vagaries of modern bourgeois atomic physicists lead them to inferences about the electron's possessing 'free will,' the replacement of matter with some sort of collection of energy-bearing waves, and to other devilish tricks."[12] Were not the ideas of these same "bourgeois" atomic physicists essential for building the bomb?

Acting in his capacity "as a physicist and not as a member of the Central Committee apparatus," Suvorov wrote a memo to Zhdanov voicing his concern. Suvorov warned Zhdanov that his statement, if published, could be interpreted as opposing modern physical theories, and physicists might thus conclude that strict adherence to the official philosophy might somehow be incompatible with their work. Suvorov also feared that philosophers and some physicists would take Zhdanov's views as implicit support from the Central Committee for attacks on quantum mechanics and relativity. As Suvorov put it, these "philosophical and physical hacks" would undermine the very theories that made the accomplishments in modern physics possible. Zhdanov's statement, if unchecked, could very well lead to a full-blown and dangerous discussion among philosophers and physicists about modern physical theories precisely when work on the bomb was of central importance to the state. Suvorov believed that such a discussion should be avoided. Zhdanov did not heed the warning and responded with only minor changes in his text. His reference to the "devilish tricks" of "bourgeois atomic physicists" remained in the published version.[13]

Suvorov's apprehensions were well founded. Quantum mechanics and relativity had sparked philosophical controversy in the West and in the Soviet Union since they were first formulated. A statement from Zhdanov on the subject was certain to exacerbate tensions over modern physical theories and dialectical materialist philosophy.

What did modern physics have to do with dialectical materialism? The question is not new, but the range of answers historians have offered illuminate what was at stake for physicists and philosophers as they prepared in the winter of 1948–1949 for the All-Union Conference of Physicists.[14] On one level, the philosophical battles in the Soviet Union were a subset of broader epistemological discussions then taking place in Western Europe and the United States. There was no consensus about the ultimate philosophical implications of recent developments in physics, and prominent physicists such as Albert Einstein, Niels Bohr, and Werner Heisenberg disagreed about the meaning of the modern physical theories they helped to develop. Among the most vexing problems was the inability of physicists to pin down concrete values for physical properties (such as position and momentum) of subatomic particles. At best, they relied on probabilistic statements. Bohr and Heisenberg also insisted that quanta did not in fact have concrete values before a measurement of them had been formulated. Furthermore, they attempted to prove mathematically that complementary measurements, such as position and momentum, could not, even theoretically, be resolved simultaneously. To some physicists, especially those associated with Bohr's "Copenhagen interpretation," these physical theories challenged conventional understandings of the nature of matter, physical causality, and, in Heisenberg's later words, "turned science away from the materialist trend it had during the nineteenth century."[15]

Vladimir Fock was the most notable Soviet participant in these debates. While he generally agreed with Bohr's notion of complementarity and Heisenberg's uncertainty, he also set out to show that they rested on dialectical materialist grounds. Simply because a physicist could not know the exact values of quanta did not mean that they did not exist in objective reality. The physicist D. I. Blokhintsev challenged Fock's interpretation, but their disagreements generally remained civil. Other physicists, including Tamm, Lev Landau, and Iakov Frenkel', preferred to stay away from philosophical debates altogether, insisting that the new physics and philosophy remain as distinct from one another as possible.[16]

There was, however, a group of physicists from MGU that formed an alliance with a particularly aggressive group of philosophers whose spokesmen included A. K. Timiriazev (the son of a famous Russian Darwinist) and A. A. Maksimov. Essentially, they argued that modern physical theories were idealist, and therefore not compatible with dialectical materialism. Despite their disagreements, Einstein and Bohr were lumped together and attacked. This group often based its arguments on *Materialism and Empiriocriticism*, Lenin's 1909 foray into the philosophy of science. In this book, Lenin harshly criticized the physicist and philosopher Ernst Mach for his positivism and idealism. Lenin was vehemently opposed to Mach's reliance on sensory experience as the basis for knowledge. Criticizing Mach, Lenin wrote: "The one 'property' of matter which is connected with the philosophy of materialism is the property of objective reality, which exists independent of our knowledge."[17] To Timiriazev and Maksimov, modern physicists were simply neo-Machists. The historian Alexander Vucinich summed up this position: "Philosophers considered it their sacred duty to wage a continual and relentless struggle against those theoretical and epistemological principles of modern physics that they regarded as contrary to dialectical materialism. Their job was to protect dialectical materialism from the eroding influences of modern scientific theory and modern philosophy of science."[18] But not all philosophers were as belligerent as Maksimov. Others sought a middle course that did not totally reject the theories of Einstein, Bohr, and Heisenberg but also did not accept the tendency in the West to emphasize the idealist implications of these theories. For them, the "incorrect" philosophical musings of Einstein and Bohr did not negate their considerable contributions to modern physics.[19] By the 1940s, the editor of *Questions of Philosophy*, B. M. Kedrov, was attempting to carve out a moderate path that accepted the role of philosophers in discussing physical theory but rejected the aggressive denunciations put forth by Maksimov, Timiriazev, and others.

Even before the philosophy meeting, accusations of idealism were heating up at MGU. In 1945, Timiriazev and others attacked Khaikin, who had taught in the Physics College since 1934, for propagating idealism and Machism in his course on mechanics. The Party Committee of the college went so far as to attempt to expel Khaikin from the Party because he refused to admit that his textbook "contradicted Marxism" and was "reactionary by its very nature." The Central Committee saw things differently. They believed that the Party Committee had unjustifiably targeted Khaikin "under the flag of dialectical materialism." As a result the Party Committee had "discredited the authority of the Party organization in the eyes of a wide circle of scientific workers." The decision to remove Khaikin from the Party was overturned.[20]

Clearly, then, Suvorov was justified in his concern that, in the aftermath of the philosophy discussion, Timiriazev, Maksimov, and others might revive their criticisms of Western physics and the "Soviet branch" of the Copenhagen interpretation. While Suvorov hoped to dissuade Zhdanov from opening a discussion on modern physical theories, other philosophers and physicists had different ideas. In his contribution to the philosophy discussion in 1947, Timiriazev set out to show that modern physics deviated from dialectical materialism because philosophers had not kept a close eye on recent developments in quantum mechanics and relativity. According to Timiriazev, these theories rested on the idealism of Mach. Mach's influence could even be discerned in the teaching of quantum mechanics at MGU. As Timiriazev wrote, "Our young [students] sincerely want to critically study modern physics from a dialectical materialist point of view, but . . . the modern textbooks are written by spokesmen for the Copenhagen school, and . . . contain statements which give clearly Machist formulas!" Timiriazev also accused Soviet journals of allowing foreign idealist philosophy to dominate their pages. He insisted that philosophers participate in the debates in modern physics in order to highlight the idealist and deviant tendencies of Soviet and Western physicists alike. To his mind, following the ideological program outlined by Zhdanov would restore dialectical materialist foundations to physics, a field that needed its own vigorous ideological discussion.[21]

Kedrov believed that Timiriazev's speech could have grave implications, and he set about preventing its publication. Kedrov wrote to Zhdanov asking that Timiriazev's attacks on Soviet physicists and discussion of modern physics be dropped from the published version of the philosophy discussion. He explained that Timiriazev's talk contained "complete slander against the front line of Soviet physicists, accusing them of being Machists" and under the direct influence of foreigners. Kedrov dismissed Timiriazev's accusations as outdated and emphasized to Zhdanov that his "excessively shrill and ungrounded attacks against Soviet science must, without question, be removed." Two days later, Zhdanov included Kedrov's memo with a note to Stalin outlining Timiriazev's argument and suggesting that the speech not be published. He was only partly successful. Though direct attacks on specific physicists were removed, a number of negative opinions about the state of Soviet physics remained.[22]

Kedrov, however, had other ways of influencing the debate. Accepting that philosophers and physicists had to discuss the philosophical foundations of physics, Kedrov took the initiative by publishing an article by the physicist M. A. Markov in the second issue of *Questions of Philosophy*. Markov's article, which came out in early 1948, set out to defend Bohr's quantum mechanics while showing its compatibility with dialectical materialism. Even though the philosophy discussion in 1947 had raised the stakes, Markov had reasons to believe that his strong defense of the Copenhagen interpretation was timely, if not entirely safe. First, his article appeared after a two-page introduction in which Academy of Sciences president Vavilov emphasized the need for philosophers and physicists to enter a dialogue about the fundamental nature of matter. Though Markov's article could not be considered definitive, Vavilov hoped that it could spark levelheaded discussion of the subject.[23] Privately, Vavilov assured Markov that his ideas had the approval of the Central Committee.[24] Furthermore, Kedrov—one of the foremost Soviet philosophers of science and a strong proponent of using the pages of the journal for discussion—backed him as well.

Strategically, Markov began his article with quotes from Marx and Engels about the physical sciences forming the basis for materialism. According to Markov, the central questions were: "Does modern physics confirm idealism?" and "Is it true that it is impossible for us to have exact knowledge of the outside world?" Only with dialectical materialism, he concluded, was it possible to answer no to each of these questions and still accept quantum mechanics. Markov distanced himself from Bohr's more philosophical positions without rejecting his interpretation of quantum mechanics. While he used the word "complementarity," he avoided referring to the "uncertainty principle" (*printsip neopredelennostei*), using instead the less ideologically troublesome phrase "correlating impreciseness" (*sootnoshenie netochnostei*). But on the central issue, Markov seemed to be implying that dialectical materialism should conform to the fundamental laws of physics.[25]

Maksimov was quick to respond, with an article in the *Literary Gazette*, declaring that Markov's thesis was an apology for Bohr, Heisenberg, and other Western idealist physicists. By publishing his ideas in a journal with a wide distribution among the Soviet intelligentsia, Maksimov ensured that the dispute would become a subject of concern for more than just philosophers and physicists. More significantly, to the chagrin of Kedrov and Vavilov, Maksimov's article introduced into the discussion the type of charged polemic that they had hoped to avoid.[26]

With the two sides thus engaged, there was little room in the philosophical journal for the kind of respectful disagreement that, for instance, divided Blokhintsev and Fock. Indeed, after Markov's article, defending Western physics became synonymous with defending the full range of views of Western physicists and thus was tantamount to subservience to the West. In part because of the reaction to Markov's article, Kedrov gradually lost control of the journal during the course of 1948. Increasingly, Maksimov and Timiriazev set the tone of the debate.

The discussion of Aleksandrov's *History of Western European Philosophy* had suggested to physicists and philosophers alike that philosophical questions concerning modern physics had to be settled. Zhdanov's speech seemed to call for criticism of Western physics, yet Markov's article in *Questions of Philosophy* and memos within the Central Committee suggested that simply dismissing Western physical theories was not acceptable. The August 1948 Agricultural Academy session shifted the ideological battleground in physics dramatically. The biology meeting not only demonstrated the Party's willingness to intervene in scientific disputes; it also revealed the extent to which questions of idealism in the natural sciences were to be taken seriously. Maksimov announced to a Party meeting at the Institute of Philosophy in late August 1948 that the Central Committee had dealt a blow to idealism in biology and could now turn its attention to crushing idealism in physics. Aleksandrov concurred and suggested that philosophers "organize a discussion of the philosophical foundations" of modern physics.[27] Evidently, Aleksandrov was not aware of the Academy of Sciences and Ministry of Education plans for holding meetings for physicists. As it turned out, over the course of the fall of 1948 philosophical disputes were added to the agenda of the physics conference at the same time that it evolved into an all-Union affair. Furthermore, philosophical disputes hinted at another layer to the conflict: Maksimov, Timiriazev, and others were prepared to question the loyalty of academy physicists. In their formulations, Soviet defenders of quantum mechanics were almost by definition unpatriotic.

As the discussions in philosophy and biology suggest, questions of Soviet patriotism became increasingly salient in the scholarly debates of the period, and the theme was taken up in physics as well. Over the course of 1948, and peaking in early 1949, charges of "cosmopolitanism" became increasingly prominent in the Soviet press, in denunciations sent to the Central Committee, and in discussions held in local Party organizations. As the term was used at the time, "cosmopolitan" was essentially the antonym of "patriot." Scientists who had lived in the West and maintained ties with foreign colleagues, and those who published their work in foreign journals, were particularly vulnerable to the epithet. A number of physicists working at the Academy of Sciences, including Tamm, Fock, Kapitsa, Landau, and Frenkel', had studied in Western Europe and the United States in the 1920s and 1930s, and their ties to the Western physics community made them susceptible to attacks from ostensibly more patriotic physicists. That many of them were Jewish only increased suspicions—in fact, the campaign against cosmopolitanism was in many ways a quasi-official endorsement of anti-Semitism. Questions of patriotism, then, were superimposed on the institutional and philosophical disputes dividing physicists. While the battle lines did not always neatly divide physicists into the same two camps on each issue (institutional affiliation, philosophical principles, relationship to foreign science, and nationality), there was a tendency for patriotic, university-based physicists to ally with philosophers critical of modern physical theories. The academy and the atomic project, on the other hand, harbored physicists anxious to defend quantum mechanics and relativity and reluctant to accept a division between Soviet and Western science.

As was the case with Aleksandrov's *History of Western European Philosophy* and Zhebrak's articles on genetics in the American journal *Science*, physicists were often held accountable for views they had expressed before the emphasis on Russian patriotism had become de rigueur. For instance, in 1943 Kapitsa gave a speech at the Academy of Sciences in which he mentioned the transcendent nature of scientific truth. The lecture, which was published in the major philosophical journal *Under the Banner of Marxism*, called for scientists, in times of peace, to commit themselves to studying fundamental truths rather than addressing technological or practical concerns. In this sense, according to Kapitsa, scientists should concentrate on the universal aspects of science and not worry about the political or economic conditions under which a particular scientific idea came about. In Kapitsa's formulation, science transcended national boundaries, since scientific truth itself did not recognize national differences. Although such a statement was acceptable during the acme of the Grand Alliance, in the late 1940s such cosmopolitan views were clearly unacceptable.[28]

Questions about the loyalty of physicists as a group surfaced before the end of the war. Of particular concern was the prevalence of Jews among elite physicists. In 1944 an Agitprop memo noted that "in many fields of specialization the percent of scientists of our primary, Russian nationality does not correspond to its historical and political importance. . . . Among the fifteen strongest theoretical physicists in the country five are Russian and Ukrainian and 10 are Jewish." For the time being, Agitprop took no action to rectify the situation.[29]

In 1946 and 1947, the Central Committee received letters from MGU physicists complaining about the lack of credit given to Russian physicists and the parallel problem of giving excessive credit to foreign scholarship. But Vavilov and others in the academy successfully rebutted charges of "kowtowing to the West," and the Science Section rejected calls for action against "unpatriotic" physicists. Symbolic action, however, was taken. In June 1947 the Politburo canceled Soviet physics journals that had been published in Western European languages. Its resolution noted that publishing work in foreign languages was not in the interest of the Soviet state and could possibly aid foreign spies. Perhaps more to the point, the resolution added that no foreign countries published scientific journals in Russian, so for the Soviet Union to publish foreign-language journals hindered the effort to "develop a spirit of Soviet patriotism among Soviet scientists."[30] Thus, an important link between the Soviet physics community and the outside world was broken.

By the spring of 1948, efforts to discredit Jewish physicists were making headway in the Central Committee. In April, Yuri Zhdanov reported to his father Andrei Zhdanov about a letter he had received from a Jewish physicist who was trying to apply the lessons of the philosophy discussion in his own field. Yuri Zhdanov summarized the letter but, in offering an interpretation of it, completely altered its meaning:

> [The author] claims that a group of physicists has formed a monopoly in their field in our country and he also recognizes among them an uncritical attitude toward theories of Western European physics. Without a doubt this hampers development of Soviet physics. However, [the author] incorrectly identifies who has established a "monopoly in physics." If you listen to him, then the Western-monopolists are Vavilov, Timiriazev, Vlasov, Akulov and Ivanenko and it is against them that he aims his attack. In reality Kapitsa, Landsberg, Landau and Leontovich are trying to form a monopoly in physics.[31]

All the monopolists named by the Jewish physicist were Russian, and with the exception of Vavilov they were all employed at MGU. The monopolists identified by Zhdanov, in contrast, were closely tied to the Academy of Sciences and except for Kapitsa were all Jewish. Even before the anticosmopolitan campaign shifted into high gear, then, the battle lines in physics were emerging. On the one side were Russians from MGU who called for the end of the academy monopoly and increased vigilance in policing cosmopolitan tendencies in physics. On the other side academy physicists, many of whom were Jewish, had close ties to the atomic project and insisted on placing their work in the context of international science. Cosmopolitanism, like idealism, splintered the Soviet physics community.

Over the fall of 1948 the president of the Academy of Sciences, Vavilov, and the minister of higher education, Kaftanov, consolidated their plans for the physics conference. In early December, they wrote a joint memo to Party secretary Malenkov clearly shifting the emphasis of the conference away from the original goal of creating dialogue between the academy and the universities. Reflecting the desire to apply the lessons of the biology discussion, their memo was concerned primarily with how the conference would address the philosophical shortcomings in physics research and teaching: "Physics is taught in higher education with complete disregard for dialectical materialism. Lenin's brilliant work *Materialism and Empiriocriticism* is still not widely used in the teaching of physics. Idealistic philosophical tendencies, which are grafted onto developments in modern physics, are not exposed or sufficiently criticized. Idealistic philosophical conclusions based on modern theoretical physics (quantum mechanics and the theory of relativity) pose a particularly serious danger to students."[32] Furthermore, the memo continued, instead of exposing trends contrary to Marxist-Leninist theory, some Soviet physicists were taking idealist positions of their own. They translated books without criticism and wrote Soviet textbooks without mention of dialectical materialism. Vavilov and Kaftanov's central concern was physical idealism. The fact that some textbooks had not emphasized the role of Russians in the development of physics was mentioned only briefly. Problems with the technical level of training in physics were also of secondary importance. They proposed that the All-Union Conference be held during the winter vacation to address these problems.[33]

The Central Committee Secretariat asked the head of Agitprop, Dmitrii Shepilov, to look into the proposal and come up with a decision. Making only minor changes to the organizational plan of the conference, Shepilov endorsed the project. As he understood it, the conference, among other things, would allow for a broad discussion of the "methodological questions" in physics, as well as provide a forum for rebutting the arguments of "bourgeois scientists who use new physics to reach idealist conclusions." On December 21, 1948, the Secretariat approved the plan and agreed to the agenda for the meeting. Education minister Kaftanov would give the opening comments, to be followed by a general speech by academy president Vavilov. Other plenary speeches would address idealism, the history of physics, the training of teachers and professors, and ways to improve the quality of textbooks and journals. In addition to the plenary sessions, the conference would also address teaching methodology, improvement of laboratory teaching, and a review of the accomplishments of modern physics.[34]

Some of these subjects clearly concerned professional matters, but other talks left plenty of room for ideological conflict. The conference's fifteen-member organizing committee (Orgkom) recognized this, and within a week of the Central Committee resolution it began to meet regularly to work out the details of what was supposed to be said, who was supposed to say it, and when. Over the next three months the Orgkom, along with other physicists and philosophers, met forty-two times in an effort to settle these questions. The assistant minister of education, A. V. Topchiev, chaired the Orgkom. Other members included Ioffe and Vul from the Academy of Sciences, Kedrov and Maksimov from the Institute of Philosophy, and Nozdrev and Sokolov from MGU's Physics College. In some respects, the Orgkom meetings reflected the original intention of bringing together academy physicists with those from MGU. The conference's structure was supposed to favor this as well. Kaftanov would chair, and Vavilov's topic, as well as his prestigious position as the president of the Academy of Sciences, would make it clear to the few hundred college physics teachers in attendance that his speech was the central focus of the conference. In fact, Vavilov was supposed to deliver his speech from the very rostrum in the House of Scientists where Lysenko had spoken in August. Topchiev and the Orgkom's charge was to make sure the conference

Figure 7. President of the Academy of Sciences, the physicist Sergei Vavilov. Courtesy of RGAKFD.

would go smoothly, by making logistical arrangements, approving the plenary speeches, and reviewing the speeches of other participants.[35]

Some of the early meetings clarified that there was room for agreement among the members. On pedagogical and patriotic issues, even academy physicists such as Vavilov, Ioffe, and Fock could agree that reform was necessary. All the Orgkom's members concurred: Soviet universities used too many translated textbooks by "bourgeois physicists," and those written by Russians did not emphasize the role of Russian scientists in the development of physics; the Cold War required scientists to praise Russian traditions and Soviet accomplishments; science was a measure of the competition between the two systems; Soviet scientists were not supposed to praise Western science; and modern physics needed to be shown as compatible with Marxist-Leninist principles.

When these generalizations turned to more specific concerns, disagreements emerged. Should quantum mechanics and relativity be removed from the curriculum altogether? Or was teaching these aspects of physics allowed, or even required, so long as it was done with a strong dose of dialectical materialist philosophy and uncompromising criticism of the idealist mistakes of Western physicists? Were those who had been reluctant to recognize the philosophical foundations of physics now supposed to be criticized?

During the Orgkom's first few meetings fundamental disagreements about the proper tone, scope, and significance of the planned conference became apparent. Vul, who had conceived of the conference back in August, accepted that a discussion of idealism was necessary but wanted charges to be directed exclusively against Western physicists. In his mind, physicists needed to present a united front, and this would not be possible if speeches attacked Frenkel', Markov, Fock, and others. Furthermore, idealism was only one item on Vul's agenda, and he believed that discussions of pedagogy and other practical matters were more central to what Soviet physics teachers needed to learn from their trip to Moscow.[36]

The MGU contingent and the philosophers insisted on an Agricultural Academy–like conference. For Maksimov, Nozdrev, Timiriazev, and others, the purpose of the conference was to purge the physics establishment of idealism and the pervasive influence of bourgeois physics. They actively criticized Fock, Markov, Frenkel', and others for their idealist mistakes. This group was faced with a fundamental problem, however. Vavilov, assigned the role of delivering the main speech, resisted attacking Soviet physicists. In fact, Vavilov delayed sending a draft of his speech to the Orgkom, which meant that others were expected to prepare their responses to Vavilov's plenary speech without actually knowing what he was planning on saying.

Confusion about the meeting's content meant that Topchiev's role as chairman became increasingly important. His position on the central question of idealism was not clear. At one point he noted: "In essence Frenkel' has been systematically conducting a struggle against materialism since 1931. Why can't we show this example to our Soviet physicists and teach them criticism and self-criticism? I think that our conference should be conducted on a level commensurate with what took place at the Agricultural Academy session."[37] At other times Topchiev seemed to temper the more aggressive speakers. In one of the early meetings he implored everyone to remain civil and, in a thinly veiled reference to atomic work, reminded them that "some of our scientists have made mistakes in the past, but now they are engaged in great, serious and high-priority work that is needed for our government. This must be taken into account."[38]

The agenda was not worked out when the Orgkom turned its attention to discussing specific talks. Agreement was pretty easy to come by when issues of pedagogy were brought up, but other subjects deteriorated into arguments about idealism. In the first few weeks, the academy physicist Vitalii Ginzburg opposed Maksimov and others anxious to discredit Soviet physicists. The Orgkom could not decide to what extent physical theories could be disentangled from their idealistic implications or to what extent individual Soviet physicists should be held responsible for their indiscretions on matters of philosophy.

Predictably, Maksimov was by far the most virulent critic of idealism in modern physics. On January 15, the speech he submitted to the Orgkom was critical of Einstein's theory of relativity as well as quantum mechanics and singled out Frenkel' as an idealist. Maksimov concluded, "Physical idealism, Machism and the like help keep physicists and natural scientists in general materially and ideologically subjugated to the American and English imperialists. . . . Physical idealism is a link that connects scientists to the hearse of capitalism."[39] Although Maksimov shared some of Lysenko's rhetorical flair, his talk was criticized not only by the academy scientists, who had arrived in full force, but also by university physicists and some of his colleagues in philosophy. They admitted that it had been an unfortunate, and even critical, mistake to have repeated the idealist conclusions of Western physicists, but they argued that Maksimov's talk did a disservice to attempts to articulate a dialectical materialist physics. Maksimov's talk was unnecessarily vindictive, and his attacks on idealism in Soviet physics placed him squarely in the minority, even among the philosophers.[40]

By far the most significant response to Maksimov's talk came from Kaftanov. The minister of education had participated only sparingly in the organizational meetings up until this point. After others had spoken their mind about what Maksimov had said, Kaftanov took the floor. Since he was the highest-ranking Party and government official in attendance, his words had particular weight. He did not shy away from the opportunity to clarify what he imagined the conference was supposed to be about. He emphasized that there were a number of heated methodological questions of great scientific significance among the philosophers and physicists, noting, "Each law of nature reflects a reality, but from each law of nature it is possible to reach different methodological conclusions." "It is clear," he continued, "that the struggle of scientific opinion about a number of questions will not be concluded at this meeting, and I believe that the struggle will not be concluded at the big conference. To give a single conclusion to these scientific, theoretical and methodological problems would be very difficult. . . . Let's take the question of the theory of relativity. As much as we discuss this question, everyone presents different opinions—opinions that I think will continue to be in contradiction with one another for a long time."[41] Clearly, this was a blow to those hoping for a conclusive Party line in physics resembling the one meted out in biology.

Furthermore, Kaftanov argued that naked criticism of modern physics should be discouraged:

It would be possible to stand at the tribune at the conference and explain that the theory of relativity is pure idealism, and say that the theory is a trick that can be refuted with materialism. But what will this accomplish? Let's say a professor from

Tashkent University attends the conference and hears our professors say that the theory of relativity is idealism. And tomorrow he has to teach and he must ask the question: should I teach the students about relativity or not? If we approach the theory of relativity like some of our comrades do, then tomorrow we need to stop teaching quantum mechanics and relativity in our institutes, not to mention that Heisenberg's theory of uncertainty would be impossible to discuss, and the Compton effect would have to be thrown out, and so on. Would this be right? Are we prepared to do this? Things are not that simple that you can take a scientific theory and turn it off, throw it in a box and say: you have done your job—now take a break. No, that is wrong. We need to show our physicists—if the philosophers were able to show them it would be useful—where in the theory of relativity science ends and idealism begins.[42]

Emphasizing his point, Kaftanov compared Einstein to Lev Tolstoy, not Weismann or Morgan:

Let's take Tolstoy. Tolstoy is a great, unsurpassed writer of Russian literature, an artist of marvelous strength. He clearly described the reality of Russia. But take him as a philosopher. Can we relate to Tolstoy the way some comrades would like to relate to Einstein: either he is materialist and therefore one of us or he is alien to us. Tolstoy is ours in the way that Lenin described when he respectfully spoke of Tolstoy as a mirror of the revolution. And at the same time his philosophy is alien to us. . . . We should critically examine all the pronouncements of Einstein, Bohr, Compton and others and those things that are useful we should not throw away. We would be savages if we refused the great discoveries and accomplishments of the exact sciences in the West. We would be savages.[43]

Kaftanov even defended Frenkel'. He noted that Frenkel' had done great scientific work and was one of the country's best scientists. And despite earlier mistakes, Frenkel' had moved closer and closer to the "right path." The role of philosophers, according to Kaftanov, was to help Frenkel', not to bring him down.[44]

Kaftanov's speech made it clear that Maksimov's extremist understanding of idealism would not be allowed to dominate the conference. But at the end of January 1949 another issue came to the fore—cosmopolitanism. On January 29, *Pravda* printed an article denouncing a group of "rootless cosmopolitan" theater critics. The next day *Culture and Life*, the organ of Agitprop, repeated the charges. Over the course of February and March 1949 the anti-cosmopolitan campaign went into high gear throughout Soviet society, with new articles appearing nearly every day in the press about yet another group of cosmopolitans. Local Party organizations held meetings to extract confessions from the "rootless cosmopolitans" (who were predominantly Jewish) in their own communities.[45]

The campaign affected the tone of the Orgkom meeting discussions. On February 2, MGU physicist Akulov presented a speech to the Orgkom in which he set about attacking "the lack of Party-mindedness" in science and promoting "Soviet patriotism." But as others on the Orgkom were quick to note, rather than speaking on behalf of any positive understanding of Soviet patriotism, Akulov had simply denounced Landau, Tamm, Ioffe, Mandel'-shtam, and others as unpatriotic and cosmopolitan. He even implied that Mandel'shtam had been a German spy. Though his speech was roundly criticized by a number of physicists and philosophers, Akulov clarified that if the conference were to be held, cosmopolitanism would be a central issue. Kapitsa, Ioffe, Frenkel', and others were clearly put on the defensive. In the atmosphere created by the *Pravda* article, charges of cosmopolitanism were potentially much more dangerous than charges of physical idealism. Eventually, Vavilov admitted that cosmopolitanism was a problem among physicists, but he still refused to name names on this point. Furthermore, the discussion of cosmopolitanism in the Orgkom meetings reaffirmed that the conference would be divisive, if held. Unlike the biology conference, it was not clear that a consensus could be formed around patriotic and materialist physics, in part because so many prominent physicists were unwilling to compromise and admit to supporting idealist theories or behaving in an unpatriotic manner. The strong response to Akulov's denunciations of Frenkel' and Mandel'shtam suggested that, like idealism, cosmopolitanism was a theme that would defy efforts to reach a consensus.[46] During March, academy physicists began showing up at the Orgkom meetings en masse. Fock, Tamm, Frenkel', and others made it clear that though they were willing to partake in some "self-criticism," on the central issues they would not back down.[47]

The draft of the conference's formal resolution, intended for publication at the end of the conference, gives some indication of the effect the conference organizers had hoped to produce. The main concern was with the uncritical acceptance of Western views and the habit of physicists to declare Soviet science "provincial."[48] Comparing Vavilov and Kaftanov's initial proposal for the conference with the draft resolution clarifies how much the purpose of the conference had changed over the winter of 1948–1949. Initially, the conference organizers had hoped to present a united front of physicists and philosophers on methodological and pedagogical questions in physics. By March 1949 questions of loyalty and patriotism outweighed the issue of physical idealism. The resolution noted that a certain sector of the Soviet physics community was "infected with the idiotic sickness of subservience to capitalist countries" and was prone to being "captivated by cosmopolitan ideas." It labeled Landau, Kapitsa, Markov, and Ioffe cosmopolitans. If the conference were held, it was clear that these physicists would be given the role of confessing their wrongdoings. But despite the draft resolution, it was not at all clear that the "guilty

physicists" would quietly accept their fate. And after three months of arguments in the Orgkom meetings, it was not clear that Topchiev had the means
to make them do so. Vavilov certainly did not have the will.

By late January 1949, shifting goals and continuing disagreements had forced
the organizers to abandon the original plan of holding the conference that
winter. On January 21, Vavilov and Kaftanov wrote to Malenkov asking that
the conference be delayed until March 21–27. They suggested that this would
give them time to improve preparations. On January 31, the Central Committee granted the request.[49] But more time for preparation could not solve the
problems. In late March (when the rescheduled conference was supposed to
be taking place), Kaftanov sent Malenkov a twenty-five-page report titled
"Major Shortcomings in the Training of Physics Cadres and Measures for Improvement." Again, in contrast with the original intent of the conference
organizers, the primary concern in the report was cosmopolitanism. Kaftanov
emphasized the following facts: Ioffe had worked in Germany and had relatives
living there; Kapitsa had lived in England for many years; Landau had numerous connections with foreign scientists; Tamm had studied in England and
Germany; and Frenkel' had been to France, England, the United States, and
Italy. To be sure, Kaftanov also mentioned problems with a generally low
appreciation for philosophy, as well as shortages of equipment and laboratory
materials. It was not clear how the conference would address these problems,
however. So Kaftanov also submitted a draft resolution to form an expert
committee to look into the situation in physics. Perhaps a committee could
succeed where discussion had failed. Significantly, in addition to Party functionaries (including Shepilov and Yuri Zhdanov) and scientific administrators
(such as Kaftanov and Vavilov), the committee was supposed to include Kurchatov, the scientific director of the Soviet atomic project.[50]

Within a few days, Kaftanov sent another memo to Malenkov asking that
the conference be postponed once more, this time until May 10. On April 9
the Secretariat of the Central Committee chose to delay the conference again,
this time indefinitely, citing a lack of preparation. Explaining the decision in
a memo to Malenkov, Shepilov wrote: "Considering the conference has not
been well prepared, the vital need to conduct deeper study of fundamental
questions, and the need to prepare concrete proposals in the field of physical
science, we propose rescheduling the [conference] for a later date. The question of when to call the conference should be decided separately."[51] In fact, the
conference was never rescheduled. In July, Agitprop clarified that Kaftanov's
proposal for an expert committee had also been shelved.[52]

Precisely why the conference was canceled has been debated among historians, in part because the documentary evidence does not precisely match the

memoirs of those who were involved.[53] Memoir literature suggests that Kurchatov told Lavrenty Beria or Stalin that the conference would interfere with the development of nuclear weapons. It is unlikely that Beria would have allowed philosophical qualms about the idealism of Western physics to hinder the atomic project. To the contrary, the extensive Soviet espionage network in the United States helped show Beria that Western physics was quite valuable, not idealist nonsense. Stalin agreed and canceled the conference, concluding, "Leave them in peace, we can always shoot them later."[54] This version rests in part on the reasonable assumption that Stalin alone had the authority to cancel the conference. Accordingly, the moves in the Secretariat and Agitprop suggested by the documentary evidence were simply an effort to give bureaucratic legitimacy to a decision that was made on high.

Another version, based primarily on archival records, suggests that bureaucratic maneuvers were responsible for the cancellation. Perhaps the first attempt to flesh out this version of events took place in 1951, when Maksimov recorded his understanding of what had happened: "Despite the considerable work conducted by the chairman A. V. Topchiev and the conference's Orgkom, the conference was not held. To be more accurate, the conference was canceled precisely because of the Orgkom's substantial work, since the Orgkom heard all the speeches and even all the proposed contributions to the conference. While the Orgkom was doing this work, it became clear that the conference could be used to strengthen the position of the idealists, the position of the cosmopolitans, and the struggle against dialectical materialism."[55] While Maksimov's opinion clearly reflects his peculiar take on the issues discussed by the Orgkom, there is likely some truth to his sense of why the conference was canceled. The lack of consensus among those preparing for the conference suggested that if it were to be held, a full range of views would likely be expressed. Whether this would strengthen the "idealists" and "cosmopolitans" as Maksimov suggested or not, it surely would have reflected the sort of disagreements among philosophers and physicists that were obviously not appropriate for an all-Union conference. The Party supported scientific conferences with the intention of settling controversies, not exacerbating them. Party leaders and conference organizers may very well have decided that it would be worse to hold a divisive meeting than not to hold one at all. Lack of adequate preparation, cited by Shepilov, corresponds in some respects to Maksimov's description of the cancellation.

Evidence of an additional bureaucratic maneuver in the Central Committee also seems to support the theory that the conference organizers and Party functionaries in Agitprop were responsible for the cancellation. Even while the organizational meetings were taking place, arrangements were being made to create a Scientific Secretariat of the Academy of Sciences. The Secretariat would report directly to the Central Committee. Topchiev would be made the main scientific secretary, a position from which he could help monitor the

activities of the academicians, thus removing one of the reasons for holding
the conference in the first place. Yuri Zhdanov would also be one of the five
secretaries, further melding the Party's scientific administrators with the academy's. If ideological issues came up in the future, Zhdanov and Topchiev
would be in a position to enforce a Party line. The Central Committee Secretariat approved the Scientific Secretariat on February 26 and the Politburo
passed the resolution on March 11, at precisely the same time that it approved
moves to postpone the conference. While circumstantial, this evidence also
suggests that the cancellation resulted, more than the memoir accounts suggest, from decisions within the Party bureaucracy.[56]

Two additional pieces of evidence related to this matter deserve to be highlighted. First, according to Yuri Zhdanov's recollections, Agitprop played the
decisive role in keeping the conference from taking place.[57] Not all memoir
evidence, then, points to Kurchatov as the decisive figure. Second, the fact
that Kurchatov was put on the membership list of the proposed committee to
look into the situation in physics suggests that he had been keeping abreast
of the situation. It seems unlikely that Kaftanov would have named Kurchatov
if Kurchatov had not already shown interest in the conference's outcome. The
first atomic test was only six months away; he surely wanted to avoid a major
attack on academy physicists. Perhaps it was at this point that Kurchatov
suggested to Beria or Stalin that working on ideological questions in physics
would distract him and others from work on the bomb. Perhaps as a result, or
perhaps just coincidentally, the committee to which he was going to be appointed never met.

In the end, historians' attempts to prove that either Kurchatov or Party
bureaucrats were responsible for the cancellation of the conference have obscured some broader points. A longer view, made possible by the minutes of
the Orgkom meetings, reveals that there was a lack of consensus on a full
range of crucial questions. The elusive consensus, rather than the specific
moves of March and April 1949, made it impossible to hold the conference.
Despite the efforts of the organizing committee, it was clear to everyone that
the All-Union Conference of Physicists would not have furthered the cause
of ideological and scientific unity. When initial efforts to solidify the script for
the conference failed, the January meeting was delayed. When March arrived
without any progress, the conference was delayed again. In short, when Kurchatov told his patrons in the state and Party that the conference should not
be held, he was only contributing to the demise of a project that was already
in trouble.

This does not mean, however, that the atomic bomb played no role in
saving physics from a fate similar to that of Soviet biology. Tacit acknowledg-
ment of the strategic importance of atomic bombs and the scientists who
created them was constantly present in the physics discussions. Without the
stunning example of the power of nuclear weapons, it is almost certain that

the number of Soviet philosophers and physicists willing to cross the line from criticizing idealist interpretations of quantum mechanics and relativity to dismissing the theories themselves would have been substantially larger. Without the bomb, it is possible that the elusive consensus about the conference's purpose would have been reached. The reputations of Einstein, Bohr, and Heisenberg in the Soviet Union were protected from a postwar fate similar to that of Weismann and Morgan because an "atomic shield" began to work as early as August 1945. Stalin accepted—at least for the time being—that academy physicists, represented by Kurchatov, had more to offer than philosophers or Party ideologues. He could afford to await the outcome of the atomic test planned for the fall of 1949 before settling the debate or purging the field.

The cancellation of the conference and the successful atomic test did not end concern about physical idealism. For the rest of the Stalin period, skirmishes persisted among physicists and philosophers. Since the conference had not been held and the Party had backed away from establishing a line for physics, the issues debated at the Orgkom were raised in journals, at Party organization meetings of institutes and universities, and at the Central Committee. *Questions of Philosophy*, now out of Kedrov's hands, published articles that lambasted Soviet scientists for uncritically assimilating the work of Bohr and Heisenberg in their textbooks. But the Central Committee clearly placed limits on the attacks against Soviet physicists. Editors at the journal *Zvezda* refused to publish an article by a scientific journalist on quantum mechanics that was critical of Fock and others. When the journalist appealed to the Central Committee for support, the Science Section leaders sided with the editors' decision. They cited Vavilov's negative review of the article as decisive and reported back to the author that they had "nothing against the publication of critical articles on the methodology of physics, but the criticism should be productive and display the tone of a scientific discussion and not just find fault with Soviet scientists."[58] Finding the proper balance, however, proved difficult. A book on the history of physics edited by Timiriazev sought to emphasize the role of Russians in physics. But it was roundly criticized in the press and in the Central Committee, which sent it out to other scholars for review.[59] When it came to questions of physics and philosophy, the Central Committee felt comfortable delegating its authority to academic experts.[60]

By far the most concerted effort to deal with the philosophy of physics after 1949 came when Kuznetsov and Maksimov set about publishing some of the speeches that had been prepared for the canceled conference. The edited volume, titled *Philosophical Questions of Modern Physics*, was first proposed in March of 1951. In addition to Maksimov, Kuznetsov, and Terletskii, others who had been active in planning the conference got involved in the volume.

Figure 8. Yuri Zhdanov, Aleksander Oparin, and Aleksander Topchiev stand while Sergei Vavilov lies in state, January 1951. Courtesy of RGAKFD.

Fock, Kedrov, Blokhintsev, and Ginzburg, among others, either wrote articles for the book or served as referees and tried to temper the content.[61] Though Vavilov died in January 1951, his speech was made the lead article. After more than a year of discussion and heated exchanges among the authors, editors, and referees, the book finally came out. It contained a number of the usual charges of idealism in quantum mechanics, while emphasizing the need for physicists to participate in the philosophical debates surrounding the field to counteract "antiscientific interpretations of the theory of relativity." Maksimov mentioned the need for philosophical consistency in science along the lines of changes in agrobiology, physiology, and other fields. He noted in the preface, "Among Soviet physicists the process of moving ahead is still too slow when compared with those events that cleared the way for progressive science in agrobiology, physiology, microbiology and cell science." Part of the purpose of the book was to energize the fight against the "so-called 'principle of complementarity' of Bohr and Heisenberg" as well as the idealist positions associated with "antiscientific interpretations of the theory of relativity."[62] The book certainly aroused anxiety among academy physicists, and it was widely discussed among them. But it did not contain many of the polemically charged slogans that had been circulating in 1949: no one was labeled cosmopolitan or accused of kowtowing, nor did the subtext of anti-Semitism arise.[63]

In the meantime, some room had been made for argument. In 1952, Fock analyzed the usefulness of D. I. Blokhintsev's idea of the microensemble in quantum mechanics in an article published in *Questions of Philosophy*. Fock was acutely aware of the dangers involved with an open discussion of philosophical issues, but he still strongly criticized Blokhintsev's theories, particularly his desire to get away from statistical and probabilistic statements about the nature of atomic matter. Fock remained close to the Copenhagen interpretation, while emphasizing that an inability to measure exact location of matter did not imply that matter did not, in fact, exist in objective reality.[64]

Fock's major contribution to the philosophical debate surrounding quantum mechanics, a rebuttal of Maksimov's position, was published in 1953 in *Questions of Philosophy* under the title "Against Ignorant Criticisms of Modern Physical Theory." The journal had not published a comprehensive defense of uncertainty and complementarity since Markov's controversial article in 1947. Fock's piece inspired historian Loren Graham to note that "one of the remarkable aspects of Fock's career, and of the history of Soviet philosophy of science, is that he was able to defend the concept of complementarity during a long period when it was officially condemned in philosophical journals."[65] Though the opinions put forth in Fock's article have been analyzed by historians of science, new archival evidence clarifies the decisive role atomic physicists played in getting the article published in the first place.

Fock decided to write the article defending quantum mechanics as a response to *Philosophical Questions of Modern Physics* and to another of Maksimov's articles, "Against Reactionary Einsteinians in Physics," published in the *Red Fleet*.[66] But it was not clear whether Fock's work would be published. In July he wrote to Malenkov complaining that Maksimov's article would "seriously hinder the development of Soviet science and technology," that Maksimov was denying the validity of the physical theories that formed the basis of all modern physics, including atomic physics, and that he was slandering Marxist-Leninist philosophy. In his letter, Fock included a copy of his own article in which he set out to show how dialectical materialism and modern physics were in fact compatible. He asked Malenkov to support its publication in a prominent journal. The issue went as far as the Central Committee Secretariat, where it was filed with no indication that the article would be published.[67]

Five months later, Fock made a second effort to get the article published, but this time went through different channels. He took advantage of the close ties Kurchatov had developed with Beria while working on the atomic project. Instead of sending the article directly to Malenkov, Fock gave it to Kurchatov, who sent it to Beria with a strong endorsement and a letter of support from eleven of the most important atomic physicists in the country, including Tamm and Landau (both future Nobel laureates), Anatolii Aleksandrov (the future president of the Academy of Sciences), and the young star of theoretical physics, Andrei Sakharov. Kurchatov wrote a short cover letter to Beria asking him to read Fock's article.[68] In a longer letter to Beria, the physicists emphasized that an unusual situation in physics had resulted from mistaken and harmful intervention by philosophers in matters of physics. The authors recognized that philosophers played an important role in the struggle between idealism and materialism but complained that some philosophers who were ignorant of the foundations of physics were now engaging in attacks on quantum mechanics and relativity.

They singled out Maksimov's article "Against Reactionary Einsteinians in Physics" as particularly dangerous and antiscientific. Maksimov's criticisms of Einstein's theory were troublesome, they claimed, because it would be impossible to solve problems of elementary particle physics or atomic power without the use of the theory of relativity. To make matters worse, Maksimov's ignorance allowed him to attack quantum theory by labeling all modern physicists "idealists." Furthermore, the authors continued, articles by other philosophers in *Questions of Philosophy* and the *Literary Gazette* indicated that this ignorance was pervasive. Finally, the letter emphasized the importance of theoretical physics to the country and the danger posed by philosophers meddling in the physicists' affairs. The scientists wanted their opinions expressed and wanted Fock's article published in a major journal to give prominent voice to their criticism of Maksimov and the other philosophers.[69]

Interestingly enough, Blokhintsev's personal comments on Fock's papers were also included in the file that was subsequently circulated among members of the Central Committee. Despite Fock's somewhat forceful criticisms of Blokhintsev's work, Blokhintsev declared that Fock's article was essentially correct. He added that Maksimov's article could cause nothing but problems for philosophy and physics and that Fock's rebuttal was timely and necessary. Blokhintsev went on to pinpoint specific problems with Fock's analysis, including his lack of sufficient attention to the contributions of Russian scientists to the development of the theory of relativity, but he still gave it an overall endorsement.[70]

Beria, evidently respecting the judgment of his "concertmaster" Kurchatov, but recognizing that questions of publication did not fall under his jurisdiction, forwarded copies of the article and the letters to Malenkov and asked that he look at them. Malenkov sent the information on to those in charge of science and propaganda (N. A. Mikhailov, Suslov, and Yuri Zhdanov) for consultation. With Beria involved, they did not risk shelving the article a second time. Zhdanov and Mikhailov responded to Malenkov by stating that the editor of the *Red Fleet* had made a mistake by publishing Maksimov's article without checking up on his background and without having the editorial skills to evaluate contributions to the heated subject of modern physical theory. They recommended that Fock's article be published in *Questions of Philosophy*. Suslov, the ideological chief, agreed.[71]

The issue was almost, but not quite, settled. Maksimov was on the editorial board of *Questions of Philosophy* and found out that Fock had allegedly told the chief editor that Beria supported the article Fock was submitting. Fock had evidently also spread this news at a speech at the Physics Institute of the Academy of Sciences. This prompted a letter from Maksimov to Beria in which he accused Fock of tricking Beria into supporting a mistaken and dangerous ideological line. Since 1948, Maksimov's letter continued, Fock had been defending the subjective and bourgeois views of Bohr, Einstein, and Heisenberg and, more recently, the philosophical views of Mandel'shtam. Maksimov implored Beria not to get mixed up with Fock and his views of Soviet science.[72] Beria ignored him.

Questions of Philosophy published Fock's article, ending the four years of philosophers' and university physicists' dominance in the journal. The exchange displays the influence of atomic physicists in shaping the public debate on philosophy and modern physics. Those who wrote on Fock's behalf came from a variety of backgrounds, but by acting together they displayed the overall unity of academy physicists both in and out of the atomic project. Not coincidentally, the physicists' letter supporting Fock's article also provides an early indication of the willingness of Sakharov to approach the state with broad professional concerns. Together, this group had considerable influence on the Party's understanding of the issues. By early 1953, Central Committee members in charge of science had lost faith in philosophers' ability to make contributions to the field. Physicists could police physics, even on philosophical issues.[73]

If we step back and recall the original positions around which the debate over the philosophical implications of modern physics formed, we can see that some scientists made concessions in order to silence the philosophers. For Landau, Tamm, Mandel'shtam, and others, the goal had been to prevent philosophers from discussing physics and to protect the field by emphasizing the philosophers' scientific ignorance. For them, the best outcome would have been to circumvent any philosophical discussion of modern physics. But the ideological atmosphere of the early 1950s forced both Tamm and Landau to endorse Fock's article, even though it represented an effort to assimilate dialectical materialism into quantum mechanics. It was a compromise they were more than willing to make when it appeared that Maksimov's new attacks would force the debate to continue. When the atomic physicists flexed their muscles on the philosophical front, the most egregious attempts to discredit modern physical theories were successfully rebutted with relatively little cost.

When Kurchatov had room to act, the results were impressive. In 1953 atomic physicists made a concerted effort to restructure the Physics College at MGU. In December the minister of medium machine building (as the atomic ministry was officially called), the minister of culture, the president of the Academy of Sciences, and the academic secretary of the Phyiscs-Mathematics Department of the academy wrote a joint letter to Malenkov and Khrushchev describing what they believed was a deplorable situation at the Physics College of MGU. The situation had been called to their attention by a group of academy and atomic physicists, including Kurchatov, Fock, and Tamm. It was clear that work at the Physics College was conducted at a low level, the teaching was poor, and the staff was generally unqualified. The letter further noted:

> An unprincipled group of workers with no scientific or pedagogical value has been running the Physics College of Moscow State University for a number of years. During this time the members of this group have driven out of the university a whole group of outstanding physicists including V. A. Fock, M. A. Leontovich, I. E. Tamm and corresponding member S. T. Konobeevskii.[74]

Further, the letter argued, that the MGU group had

> used the excuse of fighting against idealism to discredit our country's outstanding scientists, while at the same time supporting people who do not know modern physics. . . . Instead of joining work in the most prominent fields of modern physics, some members of the Physics College have spent many years conducting a fight against basic tenets of physics (the theory of relativity, quantum mechanics, etc.).[75]

The letter recommended, among other things, replacing the leadership of the Physics College with academy physicists and creating a nuclear physics department. The administrators who wrote the letter emphasized that their recommendations had emerged after consultation with many of the leading physicists in the country. They explained that although Kurchatov was not able to sign the letter, because he was out of town, "he completely agrees with our recommendations."[76]

The Central Committee quickly formed a special ad hoc committee, which included Kurchatov, to look into the matter. This committee agreed with the general conclusions of the letter and recommended that changes be made to "eliminate the existing division between academy physicists and physicists from Moscow University." In August 1954 a Central Committee resolution, "On Measures to Improve the Training of Physicists at Moscow State University," fired Akulov and Nozdrev and ordered others to change their obstructionist views about bringing prominent physicists from the academy to teach at the college. The dean was also removed and replaced by a physicist who was close to both Kurchatov and the minister of medium machine building. Soon the academicians Tamm, Leontovich, Artsimovich, Landau, and others were invited to give lectures at MGU. The postwar conflict between academy and university physicists ended with a total rout of the MGU group.[77]

Atomic and academy physicists also brought their influence to bear on the patriotic front. There was certainly some truth to Landau's quip that there was a "nuclear deterrent" preventing attacks on physicists in general and Jewish physicists in particular. In one clear example, Lev Al'tshuler continued working on atomic weapons even though he had been critical of the Party's position on classical genetics, because leading atomic physicists went directly to Beria to plead in his defense.[78] Still, in 1950 Yuri Zhdanov, along with his superior in the Central Committee hierarchy, wrote a letter to Malenkov stating that "among the theoretical physicists and the physical chemists there is a monopolistic group (L. D. Landau, M. A. Leontovich, A. N. Frumkin, Ia. I. Frenkel', V. L. Ginzburg, E. M. Lifshits, G. A. Grinberg, I. M. Frank, A. S. Kompaneets, N. S. Meiman). This group, along with its advocates—members of the Jewish nationality—has occupied all the theoretical divisions of the physical and physical chemical institutes."[79] This concern about a Jewish monopoly was buttressed by evidence of a group of mostly Jewish physicists at the Physics Institute of the Academy of Sciences that had strengthened its position by hiring close family members. This seemed to indicate that familial relations determined who worked where, rather than merit. But little action was taken to combat this so-called monopoly, even in the midst of a broad anti-Semitic campaign. Again, one can discern a subtle working of the atomic shield.

Despite the xenophobia of the period, the Central Committee measured the success of the development of Soviet physics by comparing its achievements with what was going on in the United States. Concern about the loyalty of physicists was divorced from the content of the physics they practiced. In fact, American universities became the model for how physics should be taught, how many hours students should spend in laboratories, and how many physics journals should be published.[80] A science race had begun, with the Central Committee implicitly accepting that Soviet physics still needed to "catch up with and surpass" the United States. Without the stunning example of American success in nuclear weapons, it is unlikely the Central Committee would have accepted the notion of American superiority in physics any more than it did in biology, physiology, or any other field. This recognition of a physics gap militated against actions based on anticosmopolitan declarations.

The atomic shield, however, did not always work. Perhaps the clearest example of the cosmopolitan campaign taking its toll on Soviet physics was Ioffe's removal as head of the Leningrad Physical-Technical Institute, an institution he had founded. Despite his long-standing membership in the Party and his position as the dean of Soviet physics, in the fall of 1950 Ioffe was fired because of his alleged cosmopolitan opinions and the fact that he was Jewish. Even though Ioffe had trained many of the key participants in the atomic project, including Kurchatov, he remained vulnerable to denunciations.

Officially, Ioffe stepped down because of his inability to meet the demands of the job. He wrote a letter to the presidium of the academy asking to be removed from his post: "The scale of work connected with the leadership of the Physical-Technical Institute exceeds my strength at the current time and prevents me from conducting scientific work. Therefore I appeal to you with the request to free me from my responsibilities as director of the institute, allowing me to assume responsibility for the leadership of the semiconductors group." But it is clear that the decision to remove Ioffe was made only after Kurchatov, Vavilov, and Beria had agreed to it. First, Kurchatov, along with the atomic project administrator Avramii Zaveniagin, wrote a top secret memo to Beria accepting Ioffe's removal. Then, five days later, Vavilov and Topchiev (who had been made the scientific secretary of the academy) wrote a note to Malenkov proposing that Ioffe be removed from his post because of his advanced age and the increased responsibilities of the institute.[81] In early November the Science Section of the Central Committee approved the move, noting that Beria also supported the decision. A few days later the issue was brought before the Central Committee Secretariat and then forwarded to Stalin and the Politburo, where Ioffe's removal was finalized.[82] But the real reason for his removal is clear from other Central Committee documents, which indicate that Ioffe had been fired because of his efforts to monopolize the field of physics and his antipatriotic behavior.[83] Both Kurchatov and Vavilov seem to have accepted that Ioffe was going to be removed from his post. They chose not to fight this political battle.

Kurchatov was clearly comfortable dealing both with the atomic scientists under his charge and with Beria, his boss in the government. One of his colleagues later recalled that Kurchatov was "first and foremost an 'operator,' and what's more, an operator under Stalin—and he was like a fish in water then." When asked about this statement, Kapitsa's widow later responded, "A fish in water is a happy fish and Kurchatov was not a happy man. And a fish in water would not have died so young."[84] Indeed the intensity of acting as a scientific administrator under Stalin brought serious health risks: Vavilov died in 1951 of heart disease when he was fifty-nine years old; in 1961 Kurchatov died of a blood clot in the brain when he was fifty-seven years old.[85]

Kurchatov and Vavilov were physicists, trained by academy scientists, aware of the ideological struggles taking place, and at times prepared to risk their positions to protect colleagues. With the explosion of the first atomic device in the fall of 1949, the status of atomic scientists increased considerably. With plans for H-bombs and nuclear power plants under way, the mutually beneficial relationship between physics and the Soviet government—a powerful symbol of Soviet science in the late 1950s—was already taking hold. Still, Ioffe's firing is a reminder that Kurchatov and Vavilov had to pick the places where they felt they could intervene successfully. They were both caught in the river of Stalinist politics, beating against the current only when it seemed hopeful.

Kaftanov remarked in 1949 that it was imperative for the Party to figure out where science ended and politics and philosophy began. It should come as no surprise that even in a country with a single official philosophy, definitive solutions to this problem were never reached. A wholesale attack on the intelligentsia took place at the same time that the state offered unprecedented support to science and scientists. Kapitsa and Tamm believed that the debates in the Soviet Union were not true philosophical arguments so much as ideological power struggles. But because important philosophical issues were central to debates about quantum mechanics and relativity in the West, it was hard to tell where ideological rhetoric and political intrigue ended and where philosophical conviction began. For Tamm, true debates had to take place among physicists behind closed doors and out of the public spotlight, which was tainted by the pervasive official ideology of the Party. Fock did not agree, and in the late 1950s and 1960s—long after philosophical challenges to physics had subsided—he still published articles about dialectical materialism and quantum mechanics. For him the issues were real.[86]

An exchange of letters between Tamm and Fock reveals their differences and provides a convenient epilogue to this chapter. In 1955, Tamm received an advance copy of a *Pravda* article by Fock about Einstein. In one paragraph Fock reproached the editor of a translation of Einstein's work for not criticizing the great scientist's philosophical mistakes. Tamm explained in a letter to Fock that he agreed with the editor and believed Einstein's work should be translated without comment because so much critical literature already existed about the theory of relativity in the Soviet Union that people would likely come into contact with the criticism without ever actually reading Einstein's work. "Someone with your arch-authority," he continued, "should not give all the 'Maksimovs' incentive to revive the antiscientific 'philosophical' campaign against which you fought harder than anyone else. Remove the paragraph!"[87]

Fock immediately responded that the readers of *Pravda* understood quite well that Einstein's theory of relativity was a great accomplishment. He also saw no problem in criticizing great men and argued that, in fact, there had not been critical assessments of Einstein's work published in the Soviet Union, unless Tamm was referring to Maksimov's articles. Fock exclaimed, "I would be distressed to think that you were putting my work in the same category as his!" Turning to the heart of the matter, Fock wrote, "In my review I cannot avoid criticizing Einstein's philosophical errors. To be quiet about them would be a tactical error. The one way to inoculate the theory of relativity (or quantum mechanics for that matter) with immunity against the attacks of the philosophers is to have physicists themselves recognize the philosophical mistakes of the theorists and remove them from the theory. That is what I did."[88]

These letters show the tremendous change that had taken place since Fock's article had been published in 1953 and certainly since the organizational meetings for the conference in physics in 1948. At that time, academy physicists such as Tamm and Fock had been united by a common enemy and were forced to wear "philosophical masks" (to borrow Kapitsa's phrase) to protect their scientific views. For Tamm the world of public expression remained exclusively a forum for ideological conflict, which as a physicist he hoped to avoid. Fock, however, was asserting that publicly articulated philosophical beliefs could coincide with scientific goals and even help bring them about. As the prestige of physicists grew, so too did their opportunities to contribute to public discussions.

In many ways atomic weapons placed physicists in a unique position with respect to the state. As the letters to Beria and the restructuring of the Physics College at MGU show, philosophers may have brought ideology into physics, but atomic weapons had also brought scientists into crucial contact with political authorities. In this sense, the last battle for the protection of physicists from attacks by philosophers can also be understood as one of the first efforts by physicists to assert their power within the state and Party apparatus. Perhaps this is what Andrei Sakharov had in mind when he explained that in order to understand his political views, it was necessary to understand the milieu of the scientific intelligentsia in which he lived and worked.

In the conclusion of his comprehensive study of the Soviet nuclear project, David Holloway noted that though the physics community had suffered during Stalin's reign, it remained "an island of intellectual autonomy in the totalitarian state." One of the greatest threats to this rare, and no doubt limited, autonomy occurred in the first few months of 1949, when it appeared as though the Party expected physicists and philosophers to ape Lysenko's campaign for "materialist, Michurinist biology." The cancellation of the physics conference suggested to scientists and ideologues alike that applying the lessons of biology to other fields was not necessarily a straightforward task. Lysenko's campaign had been based on a convenient conflation of personal, institutional, and, oddly enough, practical factors. In physics, the atomic bomb gave defenders of Western physical theories crucial access to patrons at the top of the Soviet system. For physicists such as Kurchatov, this increased access brought with it opportunities to control the outcomes of institutional and philosophical battles. Discussions in biology and physics, then, resulted in apparently conflicting legacies. If ideology and scholarship could be absolutely conflated, as was the case in biology, and also shown to be distinct, as was the case in physics, what would happen in other fields? In the press and in Party discussions the answer was clear: the philosophy discussion and the Agricultural Academy session remained the central referent in subsequent ideological campaigns. The aborted physics conference was officially forgotten. Yet the idea that science might occupy a realm beyond Party ideology and Marxist-Leninist philosophy remained a powerful subterranean force in future struggles to define Stalinist and Soviet science.

CHAPTER 5

"A BATTLE OF OPINIONS"

Stalin Intervenes in Linguistics

Joseph Stalin carefully concealed from the public his decisive role in determining the outcomes of the first three postwar scientific discussions. He evidently preferred to reveal his "genius" as an editor and not as a writer. That changed in 1950 when he published an article in *Pravda* in the middle of a debate in linguistics that he himself had helped foster. While his direct intervention was unprecedented, both the substance of what he wrote and the way he presented it suggested that he envisioned scientific truth as emerging from debate and discussion, not simply from Party decrees. "It is universally recognized," Stalin wrote in his article, "that no science can develop and flourish without a battle of opinions, without freedom of criticism."[1] Scholars in a full range of disciplines—not just linguistics—understood Stalin's statement as an invitation to challenge orthodox positions in their fields.

Stalin's article also brought into the open the shift from class-based categories to geopolitical categories that had been a subtle, but persistent, part of the rhetoric of the philosophy discussion in 1947, Lysenko's speech to the Agriculture Academy session in 1948, and the preparatory meetings leading up to the canceled physics conference in 1949. The "West" supplanted the "bourgeoisie" as the enemy; creating Soviet science, not proletarian science, became the goal. Until Stalin's article made these shifts explicit, linguists and even Party leaders responsible for ideology had underestimated the importance of the ideological campaign's emphasis on patriotism. Instead they had assumed that the best way to clarify the responsibilities of Soviet scientists was to rally around an ideologically rigorous, homegrown trailblazer in their field—a Michurin of their own.

For linguists, the most obvious candidate was Nikolai Marr, who had grasped the significance of uniting Marxism and linguistics in the 1920s. In a sense, he was their only option. Marr was born in Georgia in 1864 to parents who shared no common language, leading one scholar to speculate that his intellectual odyssey may, at least in part, be understood as an effort to "get mummy and daddy to speak to each other, if in some primordial past."[2] Whatever his motivation, Marr gained a reasonably good reputation as an expert on Caucasian languages. In 1909 he became a member of the Imperial Russian Academy of Sciences and soon thereafter the dean of the Department of Oriental Languages of the University of St. Petersburg. In the years before 1917 he had been working on a theory of language that was radically different from anything anyone else believed. After initially positing a link between Georgian and Semitic languages, Marr gradually discovered that nearly all languages were in fact related to one another. In describing his "New Theory of Language," Marr targeted what he labeled the "Indo-Europeanist" linguistic school that had developed in Europe and Russia since the early nineteenth century. Whereas the Indo-Europeanists regarded linguistic ties as familial (languages in a single family evolved from a common language), Marr believed similarities in language could be traced back to sounds fundamental to all languages. All primitive societies developed first a sign language and then four sound elements, which Marr identified as "sal," "ber," "yon," and "rosh." Languages were linked by their stage of development, not by a common protolanguage.[3]

Marr was one of the few members of the imperial academy who unequivocally endorsed the Revolution.[4] He easily discovered a connection between his theory and Marxism. Rather than accepting that languages evolved slowly due to migration and borrowing, Marr asserted that languages skipped from one stage of their development to another in direct correspondence to the modes of production identified by Marx. Differences in languages concealed underlying similarities predicated on economic foundations, leading Marr to the conclusion that "same-class languages from different countries—given an identical social structure—are more similar typologically than the languages of different classes within the same country, the same nation."[5]

Party bureaucrats appreciated the propagandistic value and intellectual appeal of an old-regime academician who insisted on the centrality of the economic base to his field. His emphasis on class, rather than nationality, also jibed with Soviet Marxism's internationalist rhetoric of the 1920s and early 1930s. By the time he died in 1934, Marr held a slew of administrative posts in academia, and in the USSR his theory officially became synonymous with Marxist linguistics.

Marr had more difficulty winning over his professional colleagues than the ideological establishment. Some linguists ignored Marr and worked in more traditional ways to study the peculiar grammar and syntax of specific languages. Even self-proclaimed disciples made only brief references to Marr's ideas, while branching out in their own directions. The difficulty of adapting his economic categories of analysis to the resurgence of Russocentrism in the 1930s encouraged some linguists to ignore his work. At the behest of the Party, they insisted that languages were directly associated with nations, whose existence they considered primordial. Marr's transnational conception of language development was out of sync with the emphasis on more traditional understandings of the nation, particularly by the late 1940s, when the "anti-cosmopolitan" campaign was in full swing.[6] But the Party did not renounce him or his work, leaving logical inconsistencies between the strong, Party-sponsored tendency to emphasize the glories of the Russian language and to reify national categories, and the continued equation of Marxist linguistics with Marr. Initially, Marr's antagonism toward Western European linguistic schools, insistence on the "materialist" and class-based nature of language development, and loyalty to the Soviet regime served his legacy well in the late 1940s.

Marr's disciples tried to balance reverence for his views with the demands of the postwar ideological campaign. After Marr's death, his most prominent student, I. I. Meshchaninov, headed the Academy of Science's Institute of Language and Thought (which was named after Marr) and the academy's Division on Languages and Literature. He had been elected to the Academy of Sciences in 1932, during Marr's heyday, for work that set out to support the "New Theory of Language." But Meshchaninov hardly acted like a Lysenko in linguistics. Despite the reverberations from the other scientific discussions, between 1946 and 1948 Meshchaninov maintained a moderate position vis-à-vis Marr's detractors.[7]

Another Marrist, G. P. Serdiuchenko, headed the Moscow branch of the Institute of Language and Culture. He was more eager than Meshchaninov to apply the lessons of the ideological campaign to his field and, along with the editors of the *Literary Gazette*, took the lead in publicly insisting that Marr represented the one and only path for Soviet linguistics. After Zhdanov's speech to philosophers in 1947, Serdiuchenko appeared to gather confidence in the Party's support for his campaign to increase the ideological vigor of the field. But few linguists followed his lead. When the Institute of Language and Thought and the Institute of Russian Language held a joint meeting to discuss the postwar five-year plan in linguistics, Meshchaninov did not attack opponents or insist on the infallibility of Marr.[8]

Marr's followers by no means held a monopoly on linguistics research in the USSR. In Moscow, V. V. Vinogradov had made a successful career studying non-Marrist subjects such as Russian grammar and the history of Russian literature. As the head of Moscow State University's Philology College, he refused even to pay lip service to Marr's work. In 1946 he became a member of the Academy of Sciences, and in 1947 the Ministry of Education quickly approved his *Russian Language* as a textbook. Even in the Caucasus, where Marr's work would presumably be particularly relevant, there were a number of open dissenters. The Georgian Arnold Chikobava, a specialist on the languages of the Caucasus, was stridently anti-Marrist. Just the same, he remained a prominent teacher at Tbilisi Univeristy, and in 1941 he earned membership in the Georgian Academy of Sciences. In Armenia the anti-Marrist academician G. A. Kapantsian headed the republic's linguistics institute, and the like-minded R. A. Acharian taught at Yerevan University.

While most linguists remained uncertain how to incorporate lessons from the 1947 philosophy discussion into their work, the *Literary Gazette*—under the editorial guidance of the ideologue and the ally of Lysenko Mark Mitin— aggressively attacked the field. An article titled "No, That Is Not Russian Language" accused Vinogradov of subservience to the West and "objective" and "apolitical" writing. Serdiuchenko soon added an article emphasizing the role of Lomonosov in founding the science of linguistics and noting that some Soviet scholars ignored Russian traditions in favor of Western trends. But the tone in the academic institutes remained calmer than in the *Literary Gazette*. Linguists accepted that kowtowing to the West was wrong, but they had not yet decided that Marr's "Theory of Language" represented the only road map for future Soviet success in the field.[9]

The implications of the 1948 biology discussion were more dramatic, immediate, and clear. Within weeks, linguists recognized that Lysenko's speech would have a profound effect on their work. In essence, it served as a call to arms throughout academia. Though the biology discussion had not been held at the Central Committee, *Pravda* had given it prominent coverage, suggesting that Party decisions affected scientific discourse. Lysenko's dramatic statement that the Central Committee had approved his speech, and the subsequent firing of "Mendelian-Morganists" throughout the country, also severely altered the significance and consequences of scholarly discussions. In this setting, it is no surprise that the ideological confusion characterizing linguistics up to 1948 was replaced by a well-orchestrated and directed campaign.

The Central Committee led the way. In September the Party removed the Russianist Vinogradov as dean of the Philological College at MGU and replaced him with a dedicated Marrist. By the fall of 1948, linguists were rushing to atone for their earlier tranquillity. On October 22 Meshchaninov gave a major speech—approved by Agitprop—to a group of linguists from Moscow and Leningrad. The title, "On the Situation in Linguistics," intentionally mirrored the title of Lysenko's speech at the Agricultural Academy.

The comparisons with biology were explicit. Meshchaninov began by praising the Agricultural Academy session and Lysenko's strong rebuff of "Michurin's critics." Lysenko's speech had clear implications for all Soviet scientists in their "fight against idealist and metaphysical constructs." The similarities with linguistics were particularly strong. "The science of heredity and of primordial and unchanging hereditary material is well known to historians of the development of linguistic thought. The same 'hereditary material' appears in [Wilhelm von] Humboldt's science on the '*Volksgeist*' [*dukh naroda*], according to which language is reduced to the role of simple receptacle stemming from the 'sprit of the people,' and not objective conditions." Thus Humboldt could assume the role for linguists that Mendel and Weismann played for biologists. The parallels ran deeper. "Weismann's assertion of the impossibility of directing the inheritance of an organism by a corresponding change in the conditions of that organism even finds its echoes in the basic tenets of the leading foreign linguistics schools of today . . . Saussure at the same time denies a trace of the connection between language and understanding, and does not see the influence of social factors on the development of language."[10] Since Saussure's theory viewed language as developing independently, with no role for outside influence, he was the equivalent of Morgan. In this context, Marr, of course, became the linguistic version of Michurin—a scientist who had taken a different path from the one followed by "bourgeois," "Western" scientists.

Meshchaninov outlined how Marr's work had established a scientific linguistics that contrasted with foreign theories. He also ridiculed Soviet linguists who had not fully endorsed Marr's work and legacy. Those, such as Chikobava, who accused Marr of "mechanistic materialism" had not read his later work; those, such as Vinogradov, who sought to establish a Soviet linguistics that was independent of Marr's legacy had failed to forward a distinctly Soviet, materialist, Marxist-Leninist theory of language. Finally, those who accused Marr's followers of ignoring key aspects of Marr's work, such as the "four elements," were missing Marr's larger contribution of establishing a historical and dialectical materialist science. "The value of N. Ia. Marr's work," Meshchaninov explained, "consists of expositions higher than his [specific] concepts, and lies in the decisive blow to idealist constructions of alien trends."[11] It appeared that Chikobava and Vinogradov would at least lose their jobs, much as geneticists had after Lysenko's victory.

F. P. Filin, who spoke next at the meeting, also self-consciously modeled his speech, "On Two Trends in Linguistics," after previous discussions:

> The Central Committee decrees on literature and art, A. A. Zhdanov's speech at the philosophy discussion, and T. D. Lysenko's lecture at the August Agricultural Academy session (which was approved by our Party's Central Committee) are all documents of tremendous importance. They determine the further development of Soviet science and culture, including Soviet linguistics. The lessons of the struggle between two trends in biology, and the destruction of reactionary, idealist Weismannism-Morganism, which proved sterile for practical agriculture, have a lot to teach linguists as well.[12]

Recognizing that "mechanistically" applying the lessons of philosophy and biology to linguistics would be improper, Filin nevertheless asserted that linguists were obligated to denounce the "two trends" in Soviet linguistics and unite behind a single approach. Again, the work of Marr showed the way. Filin argued that "the development of language, its history and modern condition rest on an organic connection to the history of society, the change and struggles of social formations and the class struggle." On all fronts, Marr's work provided Soviet linguists with a model for further developing their field. Marrism was rooted in Marxist-Leninist methodology, it was created during Soviet rule, it counteracted the categories of Indo-European linguists, and it provided a Soviet alternative to Western "idealist" and "metaphysical" linguistic theories. Though all this was obvious to Filin, he lamented that other Soviet linguists were not developing and defending Marr's work. Instead many Soviet linguists "conducted and conduct a secret or open battle with Marr's new theory of language." While he recognized that some scholars had responded to the battles of the early 1930s by adopting Marr's methodology, others continued to work within a tradition that was "practically in direct contradiction with Marr and based on pre-Revolutionary bourgeois-liberal Russian linguistics." Filin interpreted the previous year's articles in the *Literary Gazette* as signs for a need for increased vigilance against the newfound confidence of those—such as Vinogradov—who ignored Marr or refused to follow his methodology.[13]

A resolution by the Scientific Council of the Institute of Language and Thought and the Leningrad branch of the Institute of Russian Language repeated the themes outlined in Meshchaninov's and Filin's talks. "The meeting held at the Agricultural Academy to discuss questions of biology helps us, linguists, correctly evaluate the situation on the linguistic front, end the friendly relationship with enemy influences in our science, and broadly develop the fight for N. Ia. Marr's materialist linguistics against reactionary, idealist linguistics." Among other things, the institutes' resolution called for increased propaganda, a reexamination of textbooks and curriculum, and an emphasis on the practical application of Soviet linguistics. The philosophy discussion and the biology meeting had clearly indicated to linguists that they, too, needed to orient their work around Party decrees and ideology.[14]

More articles in the *Literary Gazette* followed the meeting in Leningrad and notified a broad sector of the intelligentsia that linguists should march in step with the ideological campaign. In October an anonymous article, "For Party-mindedness in the Science of Language and Literature," called for an Agricultural Academy–like discussion in linguistics. A month later, an article titled "Against Idealism and Kowtowing in Linguistics" reviewed the October Scientific Council meetings and called for more action against Soviet linguists who had not denounced Western trends.[15]

Throughout 1948 and 1949 scholars emphasized Marr's role in establishing a patriotic Soviet linguistics. Looking for any excuse to praise Marr, scholars in Kazan met in honor of the fourteenth anniversary of his death.[16] The promotion of Marr went hand in hand with denunciations of his critics. In March 1949, during the peak of the anticosmopolitan campaign, Party organization meetings ridiculed the Russianist Vinogradov and others for promoting pseudoscience and conducting a campaign against Marr.[17] Agitprop's organ *Culture and Life* contributed to the growing recognition that Marrism was synonymous with progressive, Soviet linguistics. With Agitprop's support, Serdiuchenko gave speeches and lectures on Marr's legacy and wrote articles for *Culture and Life* and *Pravda*.[18] Following suit, academic meetings called for increased attention to Marr's legacy and increased vigilance against "anti-Marrists" such as Vinogradov and Chikobava.

The Academy of Sciences also promoted Marr at the expense of his critics. A July 1949 decree from the academy's governing presidium, based on a report from Meshchaninov, stated that Marr was the central figure in the success of Soviet linguistics. In attacking "bourgeois," "reactionary," "racist," and "idealist" ideas, Marr helped Soviet scholars uncover a class-oriented, progressive, and materialist understanding of language. Like the articles in *Pravda*, the *Literary Gazette*, and *Culture and Life*, the Academy of Sciences harshly attacked Vinogradov, Chikobava, and others for ignoring Marr's work or for masking their "reactionary theories" behind superficial proclamations of loyalty to Marr. More general criticism was directed at everyone in the field, and particularly the linguistics institutes, for not being more determined in their promotion of Marr. In addition to calling for a more vigorous campaign in its institutes and in its publications, the academy outlined plans to celebrate Marr's eighty-fifth birthday and the fifteenth anniversary of his death. The presidium planned a publishing spree to add to Marr's growing popularity. Among the slated publications were Marr's selected and collected works, a biography of Marr, a booklet of Serdiuchenko's speeches, and articles on Marr's broad impact on Soviet science in the *Herald of the Academy of Sciences*, *Questions of History*, *Questions of Philosophy*, the *Herald of Ancient History*, and *Soviet Ethnography*.[19]

Central Committee bureaucrats trumpeted the idea that postwar Soviet politics required a renewed emphasis on Marr and his linguistic theories. In a report to Malenkov, Agitprop workers stressed their role in bringing linguistics in line with the Central Committee decrees on ideology and science. They promoted the formulation, borrowed from the biology discussion, that there existed two divergent trends in Soviet linguistics: "one of them unites the students and followers of Academician Marr and is guided in their work by the theories of Marxism-Leninism. The other trend, presented by opponents of Marr's science, attempts to save the inviable basic tenets of bourgeois linguistics." The second trend emphasized pre-Revolutionary Russian linguistic schools and revived interest in the USSR in Saussure and structuralism, leading Soviet linguistics into "the deep crisis and wasting away of contemporary bourgeois science." Vinogradov and others were mostly to blame, but Agitprop pointed out that, since Marr's death, even his followers had allowed Soviet linguistics to drift away from the questions that Marr had considered central. Furthermore, non-Marrists had published in linguistics journals without having been challenged. Agitprop leaders presented themselves as having tried, through articles in the press, to steer linguistics toward its proper path. In this way, they could excuse themselves for responsibility for the fact that the field had become invigorated only after the Agricultural Academy meeting in 1948.[20]

Attempting to coordinate the efforts of the linguistics institutes, the presidium of the Academy of Sciences, and the journals, the Agitprop workers proposed holding a small meeting of linguists at the Central Committee in November 1949. They suggested twenty participants, including Meshchaninov, Serdiuchenko, and other Marrists from institutes and organizations in Moscow, Leningrad, and the non-Russian republics. The proposal languished and the meeting was not held in the fall of 1949 or in early 1950, when Agitprop reintroduced the idea. Even without a formal meeting at the Central Committee, Agitprop kept abreast of work in linguistics and even directed linguistic research in part by participating in, and even orchestrating, meetings at the academy.[21]

By January 1950 the linguistics institutes, the Academy of Sciences, and Agitprop had agreed on the proper course of action for Soviet linguistics. The philosophy discussion, the biology discussion, and the effort to root out cosmopolitan influences in Soviet science and society all pointed to a need for greater attention to Marr's work and legacy and increased attacks on his detractors. The Party expected a united front of Soviet linguists to challenge the theories of Western linguists and to purge their own work of any explicit or implicit bourgeois influence. Marr was celebrated in publications and conferences; awards were given out in his name. Changes in the Academy of Sciences Section on Literature and Language reflected Marrism's rise. The Sector on Comparative Grammars of Indo-European Languages was renamed "General Linguistics"; the Romance, German, and Classic Languages Sector became the sector for "Foreign Languages"; and Iranian and Finno-Ugric languages were united into a single sector on "Languages of the USSR."[22] Marrist categories prevailed. In April 1950 the Academy of Sciences could report to Malenkov that its institutes had achieved some success in publishing monographs, holding scientific sessions, and preparing young cadres to further Marr's work. The problems that remained were easy to identify: there was not enough work popularizing Marr's materialist science, the excising of bourgeois influences was not yet complete, and the five-year plan did not clarify how Marr's work would be improved upon. While Meshchaninov admitted that there were aspects of Marr's work that perhaps suffered from outdated formulations or isolated errors, Marr remained the central focus of the future development of Soviet linguistics. The victory of the "New Theory of Language," though not yet complete, appeared inevitable.[23] Then Stalin got involved.

In late December 1949, Chikobava, the Georgian linguist who had been a consistent target of the Marrists, convinced the first secretary of the Georgian Central Committee, Kandid Charkviani, to send a letter to Stalin about the serious problems Marr's ideas posed. He also wanted to make Stalin aware of the unhealthy dominance Marr's disciples maintained over Soviet linguistics.[24] The letter—which Stalin's marginalia suggest he read carefully— pointed out that if all languages were class based, as Marr claimed, it became impossible to explain the use of language during primitive communism, when classes had yet to form. Charkviani also noted that Marr's idea that languages evolved through stages of development along the lines of modes of production challenged the particular linguistic and ethnic development of individual national cultures. He added that Marr had presented no credible evidence in defense of his four sound elements. Noting that Marr had argued that the main goal of Soviet linguists was to work toward a single, world language, Charkviani countered with a quotation from Stalin supporting the notion that nations and national languages would persist in the first stage of the worldwide dictatorship of the proletariat. Charkviani's complaints snowballed. In the name of bringing about a world culture, Marr had supported the imposition of Latin alphabets throughout the Caucasus, which Charkviani saw as an insult to the ancient languages of the region. In sum, Charkviani clarified that Marr's scientific mistakes had negative political repercussions: Marr was a rootless "cosmopolitan," and his work failed to contribute to either Marxism-Leninism or dialectical materialism. What had appeared revolutionary and Marxist in the 1920s now foolishly disregarded the importance of national traditions and interests. Stalin underlined a number of passages in the letter and even suggested that Marr's disciples were "blindly" following in his footsteps. He also noted a section that highlighted Serdiuchenko's role in denouncing as "bourgeois" anyone who dared disagree with Marr.[25] Charkviani included a series of Chikobava's articles along with his letter.

In many ways, the winter of 1949–1950 was a particularly inopportune time to bring scholarly disputes to Stalin's attention. After all, he spent much of his time that winter meeting with Mao Zedong and other Chinese representatives in order to work out the details of the Sino-Soviet pact. Then, in the spring of 1950, Stalin was in the midst of discussing the timing of the invasion of South Korea with Kim Il Sung. Despite such weighty distractions, Stalin made time to study the materials from Charkviani and Chikobava carefully.[26]

The police chief Lavrenty Beria may have played a critical role in turning Stalin's attention to Chikobava's work. Stalin's library contains a copy of one of Chikobava's books printed in Georgia in 1942. An inscription from the author reads, "To the Honorable Lavrenty Pavlovich Beria—here is one of the works that resulted from your reorganization of the language institute. With gratitude from the author. 16 [April], 1942." Beria—who had been first secretary of the Georgian Communist Party before being brought to Moscow to work in internal affairs—never severed his ties with that republic. Presumably, he had been a patron to Chikobava before 1942 and, after receiving the book, had passed it along to Stalin, who was also, of course, from Georgia. Perhaps Beria had created the opportunity for Stalin to hear Chikobava's complaints about the state of Soviet linguistics.[27] In any case, Stalin read Charkviani's letters and Chikobava's articles and invited them to Moscow to discuss the issues they had raised.

In early April, Chikobava accompanied Charkviani on a trip to the capital expecting to discuss their complaints "with the Party secretaries." Instead Stalin summoned them to his dacha in Kuntsevo. At the meeting, Chikobava informed Stalin that two Armenian linguists had been wrongly removed from their administrative positions as a result of the pro-Marrist crusade. Stalin immediately called the secretary of the Armenian Central Committee, A. G. Arutiunov. Stalin's end of the conversation went as follows: "You have fired Professors Acharian and Kapantsian? . . . Why? . . . There were no other reasons? . . . Comrade Arutiunov, you have acted wrongly. . . ." At which point Stalin hung up the phone. Within days Kapantsian and Acharian reclaimed their former positions in the Armenian Academy of Sciences and Yerevan University.[28]

During the meeting at the dacha, Stalin spoke with Chikobava at length, listening carefully as the linguist related his critical stance on Marr and Marrism. Toward the end of the meeting, Stalin asked Chikobava to write an article for *Pravda* on the subject. Knowing that the pro-Marr campaign in the press ran counter to his views, Chikobava asked, "Will the paper publish it?" Stalin responded, "You write it and we'll see. If it works, we'll print it."[29] A week later Chikobava sent to Stalin a draft of the article, in which he systematically developed the criticisms Charkviani had outlined in his letter. Stalin edited it line by line, at times eliminating or adding words, sentences, or paragraphs. Most significantly, he excised one of his own quotations and emphasized in his comments that languages were national in character, not class based. On May 2 Chikobava sent Stalin another draft. Again, they met to discuss Stalin's editorial comments.[30]

Meanwhile, word of Stalin's interest in linguistics made its way down the Party apparatus. Leaders in Agitprop who had led the pro-Marrist campaign began to prepare themselves for the possibility that Stalin would suddenly change course. On April 13, 1950, Central Committee secretary Suslov received a long letter seriously challenging Marr's theories and his disciples' monopoly.[31] Under normal circumstances Suslov probably would have passed the letter down the bureaucracy to Agitprop, where V. S. Kruzhkov or Yuri Zhdanov could have simply filed it without action. Instead, with Stalin now involved, Suslov forwarded the letter to Malenkov, who in turn ordered it distributed to the other members of the Secretariat for discussion at a meeting of that body.[32] If Malenkov had to make amends for his earlier support for the "monopoly of Marrists," he now at least had something on file to suggest that the matter was still up for debate. With Stalin's views perhaps unformed and certainly unknown, other Party leaders quickly lost confidence in the campaign on behalf of Marr's "Theory of Language" that they had been supporting for a year and a half.

On May 6 Stalin approved the final draft of Chikobava's article for *Pravda*, whereupon he sent it, with a note to the rest of the members of the Politburo, asking that it be published as part of a "free discussion" of the situation in Soviet linguistics. Stalin explained to his closest associates in the Party—Beria, Bulganin, Kaganovich, Malenkov, Molotov, and Khrushchev—that "Soviet linguistics is going through a difficult period. All positions of responsibility in the field of linguistics are occupied by Marr's supporters and admirers. Those that in any way disagree with Marr are removed from their posts and are prevented from speaking their mind about linguistics." Stalin added that Marr's work contained errors and expressed his hope that the discussion could help put linguistics back on a correct course. He suggested that *Pravda* dedicate a number of pages each week to the discussion.[33]

On May 9, 1950, Chikobava's article, "On Certain Questions of Soviet Linguistics," appeared on a three-page spread in *Pravda*. The editors of *Pravda* introduced the article with a note:

> In connection with the unsatisfactory state of Soviet linguistics, the editors consider it essential to organize an open discussion in *Pravda* in order, through criticism and self-criticism, to overcome stagnation in the development of Soviet linguistics and to give correct direction to further work in the field. . . . Beginning with this issue, *Pravda* will devote two pages weekly to articles discussing questions of linguistics.[34]

Chikobava opened his article with a strong attack on Marr's "elemental" analysis, noting that even Marr's supporters had recognized the erroneous nature of this aspect of Marr's work up through 1946, only to revive it in 1949. Further, Chikobava noted that other aspects of Marr's theory—such as the single, unilinear development of language—required the acceptance of his elemental theory. Since so much of Marr's work was tainted, Chikobava proposed separating Marr's productive early work on Armenian and Georgian from his later general linguistics theories. In other words, Marr could be praised as a philologist and a linguist of specific languages but not as a founder of a whole approach or methodology for linguists. Chikobava also gave Marr credit for recognizing the superstructural nature of language and emphasizing language's dependence on the economic base. Still, he accused Marr of a mechanistic approach to that relationship, noting that the structure of language (phonetics, morphology, syntax, or grammar) does not simply reflect the development of production relations. Thus, "the superstructural nature of language in N. Ia. Marr's Japhetic theory is correct only in a general sense because the specific traits of language as a superstructural category are obscured in N. Ia. Marr's teaching." Marr's theory on the class nature of language was "entirely incompatible with Marxism." Language, after all, had developed before the advent of classes. For Marx and Engels, Chikobava argued, language was first and foremost a means of communication and not a response to work, as Marr would have it. The concept of class language, furthermore, could not be reconciled with linguistic facts. Marr's assertion that "there is no common national language but there is a class language" simply could not explain the use of a single language in a single country through many economic stages of development.[35]

To replace Marr's theories, Chikobava declared that "a Marxist-Leninist history of language must be built on rigorously checked and accurately established facts." Marr could not successfully challenge Western idealist theories of language, since he never fully understood Marxism-Leninism. "If ever criticism and self-criticism were needed, it is just in this area [of general linguistics]." Establishing a Soviet linguistics based on Marxism required critical analysis of Marr's theory and the reorientation of work in the field.[36]

Some readers assumed that Chikobava's article opening the discussion was equivalent to Lysenko's speech to the Agricultural Academy session in that it set out to establish the new orthodoxy. Others were less sure. Roy Medvedev recalls that one of his friends at Leningrad University thought "Chikobava was a brave man to attack Marr's science."[37] Serdiuchenko evidently believed that the discussion had been organized in order to expose Marr's enemies and then remove them.[38] Indeed, *Pravda* gave no indication of where the Party stood on the discussion. Unlike the one-sided Agricultural Academy session, the *Pravda* series aired both sides of the issue. The following Tuesday, Meshchaninov offered a rebuttal to Chikobava's argument. For the next month and a half *Pravda* published articles attacking Marr, defending Marr, and straddling both camps. This discussion, then, contrasts sharply with the discussion in philosophy, where criticism of Aleksandrov was mandatory, and the discussion in biology, where Lysenko's opponents were in a distinct minority. Each Tuesday over the course of the spring of 1950, readers of *Pravda* were exposed to an academic battle with no obvious, predetermined victors.

The uncertainty of the outcome was not the only peculiar feature of the linguistics discussion. Unlike other discussions, participation was limited— with one key exception— to trained experts, in this case linguists. Philosophers were not involved. The discussion took place on the pages of *Pravda*

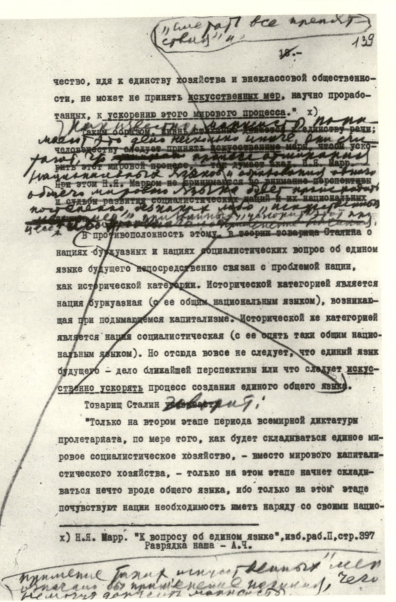

Figure 9. In early May 1950 Stalin read and edited Arnold Chikobava's article "On Some Questions of Soviet Linguistics." After approving the final version, Stalin circulated it to the Politburo and arranged for it to be published in *Pravda* to open a "discussion" of linguistics. Courtesy of RGASPI.

and not at a specific place, either the Central Committee or the House of Scientists. There were no physical clues from Party leaders in attendance about what positions were deemed correct, or whose views represented the Party's opinion. The discussion took place publicly over the course of nearly two months, gradually raising the tension associated with the discussion's outcome. While it was obvious that the Party was directly concerned with the discussion (why else dedicate such a large portion of *Pravda* each week to it?), the Central Committee's position was unclear. In the months leading up to the discussion, Marr seemed to be in ascendancy. But the first article presented a message that was in direct contradiction to the apparent Party line.

Those responsible for the Party line tried to readjust on the move. On May 13, only four days after Chikobava's article, Agitprop's Yuri Zhdanov and Kruzhkov proposed to Malenkov that Serdiuchenko be removed from his position as head of the Moscow branch of the Institute of Language and Thought, as well as from his administrative posts in the Ministry of Education and the Academy of Pedagogical Sciences. They reported that Serdiuchenko was not an accomplished linguist but instead was a "careerist, and a man who uses administrative methods to run science and to persecute scholars that disagree with his views." Only months before, Agitprop had taken credit for the publication of Serdiuchenko's articles in *Culture and Life* and *Pravda*. Now it distanced itself from his "monopolistic" behavior, if not his specific ideas. A few days later they revealed why. In a meeting with Serdiuchenko, the two Agitprop officials explained to him that he had prevented criticism of Marr and that his speech in Armenia was responsible for the firing of the two linguists, Kapantsian and Acharian, whom Stalin subsequently reinstated to their academic posts. Obviously, Kruzhkov and Zhdanov knew that the Armenians were once again in favor and that someone had to take the fall for their having been fired in the first place. They proposed the logical candidate, Serdiuchenko.[39]

Hedging their bets further, Zhdanov and Kruzhkov drafted a resolution reinstating the Russianist Vinogradov to the post of dean of MGU's Philology College.[40] This draft resolution indicates that on the eve of the discussion the Marrists no longer had the unqualified support of the Central Committee apparatus.[41] Still, despite the backstage activity, Meshchaninov remained the most powerful linguist in the country, and no actions had been taken against him. Indeed, with the exception of Serdiuchenko, the other vocal Marr supporters, including Meshchaninov and Filin, participated in the discussion in *Pravda*.

Rather than establishing the new orthodoxy, then, Chikobava's May 9 *Pravda* article was the first salvo in a debate on the future of Soviet linguistics. A week later, Meshchaninov quickly came to Marr's defense in *Pravda* with an article titled "For a Creative Development of Academician N. Ia. Marr's Heritage." Meshchaninov disagreed with Chikobava that Soviet linguists needed to begin developing a theory of language from scratch. To his mind, the problems in linguistics stemmed from a lack of attention to practical tasks of developing the culture and language of the peoples of the Soviet Union and not with any fundamental problem with Marr's work or legacy. Theoretical problems in Soviet linguistics stemmed from the persistent influence of "bourgeois" theories of the "Anglo-American" bloc. Marr's work placed linguistics on a materialist basis, stressing the social context of language development. Since language is a superstructural phenomenon, as Marr pointed out, it follows that its grammar and structure cannot be studied independently of the social context in which it was created. Perhaps hoping that noting parallels with the situation in biology would still be to his advantage, Meshchaninov emphasized that Marr had argued against those who were dismissive of social, or environmental, forces in language development. While Meshchaninov admitted that Marr's work contained some "erroneous and still uncorrected theses," he insisted that Marr's emphasis on Marxism and materialism made his ideas the basic starting point for all further research in Soviet linguistics. Finally, he accused Chikobava of affirming the very "bourgeois" linguistic theories that Marr was rightfully attempting to remove from Soviet science. Chikobava's emphasis on "protolanguage" echoed the comparative method of Western linguists. In conclusion, he called for the increased publication of Marr's papers, an invigorated effort to develop those aspects of Marr's theory that remained unclear, and the use of Marr's work to "assure the further development of materialist linguistics and . . . the critique of the basic conceptions of bourgeois linguistics." This was essentially a reiteration of the position linguists had reached before the onset of the discussion.[42]

The articles by Chikobava and Meshchaninov established the basic contours of the discussion in *Pravda* as it unfolded through June 20. Following their lead, subsequent articles concentrated on a number of points, some leading to general agreement, others clarifying strong disagreements. Marr's defenders hoped that his work would be either equated with Marxist linguistics or declared the basis for further development of Soviet work in the field. That is, they hoped to continue the approach established in the last year by the Academy of Sciences and in the popular press. Their opponents sought either to open the field to competing approaches or, better yet, to reject Marr's theory outright. Either explicitly or implicitly, each of the articles addressed the question of the relationship of Marr's work to the Party's mandate and to the future of Soviet linguistics.

Nearly all the participants accepted the validity of the editor's note in *Pravda* that opened the discussion: "criticism and self-criticism" was necessary to end the "unsatisfactory state of Soviet linguistics." To Marr's supporters, actions by the Academy of Sciences presidium and linguistics institutions over the course of 1949 and early 1950 had begun to rectify the inadequate development of the field.[43] For them the solution to problems of productivity and theoretical advancement would come with increased attention to Marr's legacy. For Marr's detractors, however, there was more incentive to talk about a monopoly and lack of criticism in their field. Still, they did so in relatively muted terms, since there had yet to be official word from the Central Committee that Meshchaninov—the presumed leader of the monopoly—was in disfavor.

G. D. Sanzheev, a Moscow professor who was relatively neutral on the question of Marr's legacy, mouthed familiar slogans from the period in pinpointing what he felt was at the root of the problems in the discipline. Indeed, his critique of his field could have been applied more broadly to the state of Soviet science in general:

> The basic and main cause of this stagnation is that all criticism and self-criticism is utterly lacking among us linguists. Those "discussions" which have taken place in recent years . . . have either been too general in character or have been limited to the "analysis" of mistakes of individual linguists. In these "discussions" fundamental problems of language were not discussed on their merits and not even raised, while thorny issues were avoided. In short, these "discussions" were more frequently carried through for form's sake and in order that everything might remain as before until the next statement in the papers: if no articles were forthcoming, that meant there was nothing to discuss, and everything was "in order."[44]

In his article, Vinogradov reminded *Pravda*'s readers who was responsible for the dogmatic interpretations in the first place: "the followers of the great Soviet scholar began to transform his theories into dogma and to view Academician Marr as a compulsory and unavoidable bridge between all Soviet linguists and the classics of Marxism-Leninism, even on those questions that Academician N. Ia. Marr himself never addressed." According to Vinogradov, the problem was not so much with Marr's often erroneous theories but with the uncritical attitude toward those theories promoted by his students and followers.[45] Kapantsian, recently reinstated as director of the Armenian Institute of Language, also thought the only way to end stagnation was to critically assess Marr's general linguistic theories. With the Marrists in mind, he quoted Stalin, who wrote, "Science is called science precisely because it knows no fetishes, is not afraid to combat the obsolete, the old, and it listens carefully to the voice of experience and practice."[46] Scholars on all sides of the battle lines could agree that linguistics had stagnated and needed an infusion of criticism. The question became to what extent Marr's work would remain associated with Marxism and Soviet linguistics, and to what extent it would be rejected as a result of the discussion.

Perhaps the least controversial aspect of Marr's contribution to Soviet linguistics was his insistence on language as a superstructural phenomenon. All sides praised Marr's desire to develop a linguistic theory that would make a viable alternative to structuralism in particular and Western linguistics in general. Even those articles that were extremely critical of Marr accepted the general goals of his work as unimpeachable parts of Soviet doctrine. Vinogradov recognized the great role Marr played by waging a "fierce and uncompromising battle with [bourgeois-idealist science] in the name of materialist linguistics." Although Kapantsian found Marr's approach "unhistorical and cosmopolitan," he admitted that "the contributions of Academician Marr consist mainly in that he posed linguistic problems materialistically in his approach to language as part of the cultural superstructure and particularly in his critique of the idealist position of the Indo-Europeanists."[47]

Marr's detractors ridiculed his theories about the origin of languages, particularly his "elemental analysis" and his insistence on the class basis of all language. Considering the degree to which the "four elements" analysis had been brushed aside even by the Marrists from 1934 to 1948, it is not surprising that its recent revival was vulnerable to criticism. Serebrennikov summed up what many other participants argued, namely that "the four-element theory was from the very beginning built literally on thin air." Marr offered no explanation for why there were specifically four elements, nor did he adequately explain the methodology he had used to come up with them. Chikobava, Serebrennikov, and even those less critical of Marr's theories in general deemed the "four element" analysis useless. Furthermore, Marr relied on his "elemental analysis" to explain the development of language over time, through a process of hybridization. In his view all language derived, in some way, from these elements. Again Serebrennikov tersely stated, "What is erroneous from the very beginning can only lead to greater error in the future." Problems with the "elemental analysis" brought Marr's whole theoretical approach into question.[48]

Marr's supporters mounted what might be called a "bend but don't break" defense. Rather than claiming that the "four elements" were indeed the building blocks of all language, they tried to limit the broader damage that could be associated with this admittedly erroneous idea. Nikolai Chemodanov, who had replaced Vinogradov at MGU, acknowledged that "no Soviet linguist used the technique of element analysis after N. Ia. Marr's death." Since so much time had elapsed since the birth of language, he accepted that Marr's conjectures about the four original elements remained open to debate. But having conceded the point, Chemodanov insisted that "the paleontologic investigations of N. Ia. Marr, based on element analysis, have revealed such incontrovertible semantic regularities as the functional semantics of the word. Consequently the principle of element analysis cannot be simply eliminated from science." The important thing, from Chemodanov's perspective, was that Marr's broader claims were viable. The details may have been wrong, but the lessons drawn from them were correct.[49]

A similar process of conceding a point in order to defend a broader principle took place in relation to Marr's claim that all language was class based. A number of scholars followed Chikobava's lead and pointed out that Marr's concept of class was incompatible with Marxism. They noted that Stalin had stated in "On Dialectical and Historical Materialism" that in the primitive commune there were no classes. Yet language certainly existed at that time. Again, this called into question Marr's insistence on the class character of language. Even Marr's supporters had to admit that his definition of "class" was simply wrong.[50]

Both the "four elements" and Marr's concept of "class" raised potentially devastating questions about his theory of the origins of languages. While some sought to exploit these weaknesses in Marr's work to challenge his whole "New Theory of Language," others sought to limit the damage and forge a path that could save Marr's more general points from ridicule. The dynamic recurred in debates about Marr's notion of the development of language over time and his identification of distinct stages in language development. San-zheev defended Marr's position, noting that it was in agreement with "materialist dialectics": "language develops not only by evolution, i.e., by quantitative changes in various aspects and facets of language, but also by revolution, by skips and mutations, i.e., by the transition of this language from one qualitative state into another . . . all of which in the long run is determined by the corresponding changes in the means of production of a given society." As with production relations, so too language developed through periods of evolution leading to explosive revolutionary breaks. He argued that since Chikobava saw language developing without such breaks, he was in direct contradiction with Marxist-Leninist teaching. Serebrennikov, on the other hand, saw this aspect of Marr's theory as a complete departure from Marxism. He questioned the empirical foundation for Marr's early historical stages, declaring, "the artificiality and abstractness of such an outline, utterly unrelated to any concrete history, is completely obvious." Another participant sought a middle ground between strict adherence to stages and their dismissal, noting that "towards the end of his life N. Ia. Marr used the term 'development by stages' much more warily . . . far from everything in his early works should be taken literally," since Marr had "deliberately allowed himself peculiar 'exaggerations' " when it came to promoting his methodology. The discussion seemed to be leading toward a compromise in which Marr's work was still declared valuable, even as his errors were recognized. His work might remain a diminished, but still significant, anchor for Soviet linguistics.[51]

Vinogradov, for one, was insistent that Marxist-Leninist linguistics not be "locked within the framework of the so-called new theory of language." Rather than outlining the areas where Marr was wrong, Vinogradov concentrated specifically on those areas in which Marr's work appeared irrelevant. Marr concentrated primarily on the origins and development of language, while Vinogradov was concerned with the way language worked in the present. This, he argued, was of much greater practical concern for Soviet linguists since they had responsibility for language codification and teaching in the USSR. Because Marr's work was useless for the current demands of Soviet culture, it was wrong to try to expand Marr's influence into new areas or to equate his work with Soviet linguistics in general. In fact, Marr was an obstacle to the study of contemporary languages. Vinogradov called for a positive reevaluation of the pre-Revolutionary heritage of Russian linguistics, particularly A. A. Shakmatov's work on Russian grammar and syntax. Citing Andrei Zhdanov's emphasis at the philosophy discussion on the value of studying pre-Marxist Russian philosophy as a precedent, Vinogradov emphasized that Russian linguistic science was a central concern of Soviet science.

Vinogradov's rejection of the utility of Marr's ideas for the study of modern languages did not mean a naïve return to the comparative-historical theories of "bourgeois" linguists. He also understood the notion of "protolanguages" as metaphysical and overly simplified. Still, a comparative analysis of the internal processes of languages revealed that there was a kinship of languages that was "material" and not "metaphysical." The trick was to reject the notion of protolanguages without repeating the "largely erroneous history-by-stages and ethnogenetic inventions of Academician N. Ia. Marr." Marr's paleontological approach was not fruitful, and the time had come for Soviet linguists to develop the comparative-historical approach.[52]

Others emphasized similar points, noting that protolanguages were not essential to comparative-historical approaches and that, far from being "formalistic," comparative-historical linguistics was useful and practical for the "living languages of the USSR." Rather than being understood as class based, language was thought of as a form of communication related to the nation. As one participant put it, with Marr "the study of the connections between related languages is turned over as a monopoly to bourgeois linguistics. This is where timidity in the face of theory and un-Bolshevik fear of difficulties lead us."[53]

Marr's followers did not accept the introduction of an alternative methodology for Soviet linguistics without a fight. They emphasized the importance of archaeology, history, ethnography, and anthropology—fields they credited Marr with bringing to the study of language—for linguistics. For them, the "comparative-historical approach" was automatically linked to "bourgeois idealism" and could not provide the starting point for a Soviet Marxist linguistics. Filin was the most aggressive in his attack on any resurgence of "bourgeois linguistics," denouncing as un-Marxist and idealist the "prolonged evolutionary process" and the presumption of a common protolanguage depicted by comparativists. By looking only at the internal structures of language, the comparative method treated languages like "some kind of biological organism" that evolved without external influence. The method, he declared, ignored vocabulary and the "qualitative changes in the very content of language." Like genetics, "formal-comparative" linguistics was associated with the "race theory of the Hitlerite 'masters' who held that the 'highly cultured' 'proto-language' was preserved in its purity and inviolability only in the German language."[54]

In the debate over linguistics methodology, it is easy to discern echoes of the scientific discussions in philosophy, biology, and physics. Like the philosophers, linguists felt the pressure to integrate pre-Revolutionary Russian science into their work. It is not surprising that Vinogradov quoted Zhdanov precisely on this point. The postwar period invited a reexamination of the Russian scientific heritage that in turn often meant the decline of the aggressively class-based approaches to scholarship forged during the Soviet 1920s and early 1930s. The Marrists were more likely to cite the biology discussion, since Lysenko provided an example of a Soviet scientist confidently defying Western theories while proudly presenting his own work as Soviet and class based. Finally, Vinogradov and others' insistence on using foreign theories while excising "bourgeois" or "idealist" elements is reminiscent of physicists who were anxious to sift through Western work on quantum mechanics and relativity in order to separate "objective science" from philosophical digressions. It was still unclear, however, how the linguistics discussion would conclude. Would the academic stagnation in the field continue, despite the effort, as it had in philosophy? Would the Central Committee designate a Lysenko-like victor by endorsing one scholar's contribution to the discussion? Or would the central disagreements remain unresolved, as had happened the year before in physics? A review of the discussion up to June 20 suggests that any one of these outcomes was possible.

Joseph Stalin's article "On Marxism in Linguistics," published on June 20 as part of the linguistics discussion in *Pravda*, completely altered the field of Soviet linguistics. The article also brought about a monumental, but ambiguous, shift in Soviet efforts to understand the relationship between Party ideology and knowledge. Stalin's participation became the focal point of all subsequent discussions of science during his lifetime. Linguistics had found its Lysenko. And Soviet science had found a new lodestar by which to navigate through the sea of Soviet ideology. Yet, the specific implications of Stalin's ideas for other fields were never entirely clear, leaving as much confusion as clarity in their wake.

At some point in early 1950 Stalin began to prepare his own statement on the situation in Soviet linguistics. It is possible that Stalin knew he was going to publish an article of his own even before he ordered *Pravda* to open up its pages to discussion. It is also possible that in the process of the discussion Stalin realized that the anti-Marrists would not necessarily prevail without his support. While the extant archival materials leave considerable room for speculation, it is clear that Stalin made learning about linguistics a priority of sorts. The collection of books in his library suggests that he had some basic works relevant to linguistics, particularly those of Marr and his disciples. Comments in the margins suggest that Stalin had read them, but perhaps prior to his discussions with Chikobava. Apparently, the *Great Soviet Encyclopedia* provided him with his most important source. Pencil markings suggest that he read the sections on Marr, linguistics, and "Japhetic theory."[55] But it did not take much systematic immersion in the field for Stalin to recognize what he believed to be Marr's mistakes.

Stalin began his article with an explanation for his participation. He claimed that a group of young comrades had asked his opinion about the linguistics discussion and he reasoned that his knowledge of Marxism gave him the authority to speak on the subject. As he put it, "I am not a linguist . . . but as to Marxism in linguistics, as well as in other social sciences, this is a subject with which I have a direct connection."[56] His somewhat self-deprecating tone was matched by an understated presentation. Like the other articles in the discussion, Stalin's contribution appeared on pages three and four, with little fanfare, and with "J. Stalin" printed simply at the bottom of the last section. Stalin evidently wanted his role in the discussion to be as an expert in Marxism rather than as the general secretary of the Communist Party. Needless to say, he could rest assured that his first direct, public foray into postwar academic debates would be read as the most significant Soviet ideological statement in years.

The article was organized as a series of questions, presumably from "young comrades," followed by Stalin's responses.[57] The first question was: "Is it true that language is a superstructure over the base?" The superstructural nature of language had been one of the few points of agreement in the discussion, and it had been the one aspect of Marr's theory that even his critics accepted as true. Stalin's response: "No, it is not true." Language was neither part of the economic base nor part of the political or cultural superstructure. He reasoned that language served all classes and all societies, regardless of economic systems. In fact, the Russian language had "remained basically the same" despite radically different superstructures associated with the periods of feudalism, capitalism, and socialism. Russians could still read Pushkin's poetry, written in the feudal period. So Stalin asserted that any talk of radical breaks in language corresponding to changes in productive forces was nonsense.[58]

Stalin then responded to a question about the claim that all languages are class based. He began with an argument brought up in the course of the discussion: in a society without classes there can be no class-based language. Since the primitive commune preceded class formation, the notion that all language was class language was clearly wrong. He dismissed the notion of "language of the bourgeoisie" and "language of the proletariat," which had been used to replace the concept of national languages with one of class languages. Dialects and jargons associated with certain classes existed, but this did not mean that languages on the whole were class based. Languages were used by societies in all stages of their economic development; thus language served all classes and in fact was indifferent to class. Stalin believed that the misconception of class-based languages grew out of misreading statements by Marx and Lenin about culture.[59]

Stalin emphasized that language was a means of communication, directly related to thought, that had existed in society throughout history. Vocabulary formed the building blocks of language and was organized according to grammatical rules. Vocabularies gradually developed over time, while grammatical structures changed much more slowly. This explained how languages had been able to serve society during many different eras. He dismissed Marr's notion of the hybridization of two languages, arguing that when two peoples or societies blended, one or the other language's structural rules won out.

Stalin also addressed the usefulness of "discussion" as a means for improving linguistics. He praised the exchange in *Pravda*, seeing great benefit in its exposure of a "regime in the center and in the republics . . . not typical of science and men of science." Perhaps thinking of Kapantsian and Acharian, Stalin noted: "The slightest criticism of the state of affairs in Soviet linguistics, even the most timid attempts to criticize the so-called 'new theory' in linguistics was persecuted and stifled by the directors of linguistic circles. Valuable scholars and research workers in linguistics were removed from their positions and reduced in status for criticism of the heritage of N. Ia. Marr and for the slightest disapproval of his teaching." He noted that others rose in prominence simply for praising Marr's work unconditionally. This had a negative effect on teaching and research.[60] .

What would this mean for Meshchaninov and others responsible for this situation? Stalin likened Meshchaninov's monopoly of the field to the policies of Arakcheev, a minister associated with the harsh measures of Alexander I's reign. The discussion, Stalin emphasized, was useful precisely because it was helping to crush this Arakcheev regime in science. In a sentence that signaled a need for greater openness in all fields, Stalin stated: "It is universally recognized that no science can develop and flourish without a battle of opinions, without freedom of criticism." The "unceremonious trampling" of this rule had led to the stagnation in linguistics. Still, Stalin suggested that these scholarly errors were not the same as political crimes: "Were I not convinced of the honesty of Comrade Meshchaninov and other linguists I should say that such behavior was equivalent to wrecking." This line might very well have saved Meshchaninov and others from arrest.[61]

For Stalin—and thus subsequently for everyone else—the solution was not to reexamine Marr's work. Marr was "merely a simplifier and vulgarizer of Marxism" and was useless for further efforts to develop Marxist Soviet linguistics. Stalin added, "Save us from the Marxism of N. Ia. Marr," which had introduced into linguistics the un-Marxist notions of language as superstructure and the class nature of language. Marr would have no future in Soviet linguistics: "he confused himself; he confused linguistics. It is impossible to develop linguistics on the basis of a wrong formula which contradicts the entire history of people and languages." To top it off, Marr had "introduced into linguistics an immodest, boastful, arrogant tone, not characteristic of Marxism and leading to the wholesale and irresponsible rejection of everything in linguistics before N. Ia. Marr." Stalin suggested that the comparative historical method, along with the liquidation of the Arakcheev regime, constituted the starting point for the revitalization of Soviet linguistics.[62]

Though the outcome had clearly been settled, the discussion in *Pravda* did not end with Stalin's article. The next Tuesday the drubbing of Marr and Marr's followers continued hand in hand with universal praise for Stalin's article. The discussion officially came to a close one week later, on July 4, with a short article by Stalin clarifying some of his points, as well as a programmatic article by Vinogradov, and self-critical articles by Meshchaninov, Chemodanov, and others.[63] Stalin's second article, presented as a response to a Comrade Krashennikova, explained that language was neither part of the base nor superstructure nor some "intermediate phenomenon."[64] He also clarified that the relationship between thought and language had to be reconsidered from a position that avoided the Marrist "swamp of idealism" resulting from the separation of the two. Elaborating on his earlier article, Stalin emphasized that class did influence language through specific words and expressions. After repeating that Marr made "flagrant mistakes when he introduced . . . elements of Marxism in a perverted form," Stalin allowed that not all of Marr's work on individual languages needed to be rejected. Still, he left little room for any rehabilitation of Marr, concluding, "N. Ia. Marr and his closest colleagues introduced theoretical confusion into linguistics. To do away with stagnation, both [the Arakcheev regime and the theoretical confusion] must be abolished." Stalin emphasized the connection between rectifying the situation in the field and a struggle between Soviet and foreign science: "The liquidation of ulcers will cure Soviet linguistics, will lead it onto a broad path and will enable Soviet linguistics to occupy the first place in world linguistics."[65]

After Stalin's first contribution to *Pravda*, subsequent articles printed as part of the discussion, and articles in a variety of other publications, praised Stalin's insights and leadership. Historians, philosophers, physiologists, biologists, economists, and archaeologists, as well as a full range of nonscholars, wrote letters to *Pravda*, *Culture and Life*, the Central Committee, and Stalin personally. Stalin had not delivered a major theoretical statement in years, which only increased the significance given to his June 20 article on linguistics. His work was lauded as a "triumph for Soviet science" and "a new and important stage in the development of science." One example of the purple prose that abounded gives a sense of the adulation: "Words cannot express our deep gratitude to our dear teacher, the great and wise Stalin, for his work on questions of linguistics. A bright feeling of joy permeated everyone after the appearance of J. V. Stalin's classic article on *Pravda*'s discussion page." (Some exaggeration is obvious; did those linguists and scientific administrators who had hitched their reputation to Marr's star really experience a "bright feeling of joy" upon reading Stalin's article?) Meshchaninov and Chemodanov admitted to errors in their work and recognized the "invaluable aid" of Stalin's article and thanked him for his "fatherly assistance." Triumphantly, *Pravda*'s editors declared the discussion closed, remarking: "The great and vital principle of the development of all Soviet science is contained in J. V. Stalin's words: 'no science can develop and flourish without a battle of opinions and without freedom of criticism.' "[66]

Stalin's intervention into the linguistics discussion elicited a series of Central Committee decisions to reorganize Soviet linguistics. Upon reading Stalin's article, Serdiuchenko evidently reported to Chemodanov, "Things are clear. I better look for openings somewhere in Riazan or Voronezh."[67] While administrative changes were indeed made, linguists who had supported Marr were not subject to arrest and, with few exceptions, were not forced to leave Moscow and Leningrad. That Stalin was "convinced of the honesty of Comrade Meshchaninov" apparently helped. The historian Alpatov reports that only two linguists left Leningrad; they lived for a few years in Ivanova and in Vil'nius. Given how easily political mistakes could lead to prison sentences during that period, this hardly seems devastating.[68] Marrists generally kept their teaching and research positions but lost their administrative posts. Meshchaninov, for instance, remained a full member of the Academy of Sciences and, for a while, head of a research unit, but in July 1950 the Party removed him as director of the Institute of Language and Thought. Serdiuchenko and Filin lost their administrative jobs as well. Central Committee decrees merged the Institute of Russian Language and the Institute of Language and Thought into a single Institute of Linguistics. Vinogradov became director and also assumed Meshchaninov's position as the head of the Academy of Sciences Division of Literature and Language. The expert commission for linguistics at the Ministry of Education completely purged itself of Marrists.[69]

Administrative changes reflected a renewed emphasis on the importance of the Russian language. Vinogradov was not the only Russian specialist to be appointed to a position of authority. The Party made S. D. Nikiforov, a specialist in Russian and Church Slavonic, an assistant director of the new institute; P. Ia. Chernykh, a specialist in the history of Russian, an editor of the academy's literature and language journal; and V. P. Sukhotin, who studied the syntax of nineteenth-century Russian, a scientific secretary of the Academy of Sciences presidium. Agitprop also appointed a Russian-language expert to its staff.[70] These decisions suggest that forging a new direction in linguistics meant emphasizing the history of the Russian language, its grammar, syntax, and relationship to other Slavic languages. This national emphasis, as opposed to Marr's transnational linguistic theory, fit with the patriotic fervor of the era.

Agitprop also quickly set about making changes in the structure and personnel of the academy's language institutes and the Ministry of Education's philology and linguistics departments. It commissioned new textbooks and course curricula to replace the outdated ones and gave special emphasis to promoting younger scholars and the study of Slavic languages. Plans to publish a special pamphlet containing all the articles from the discussion were scratched, but a special collection of essays about the meaning of Stalin's articles for linguistics was slated for publication in 1951. Stalin's articles, gathered into a single booklet, were published in an initial run of 500,000 copies. Agitprop also took responsibility for sending a group of scholars around the USSR to evaluate the state of linguistics research outside Moscow and Leningrad. Agitprop outlined thirty-one specific tasks that the Ministry of Education of the USSR, the Ministry of Enlightenment of the RSFSR, and the Academy of Sciences had to fulfill before the end of the summer of 1950.[71] Kruzhkov and Yuri Zhdanov, along with other workers from the Science Section, attended academy meetings on linguistics and closely monitored progress in the field.[72] In order to keep tabs on the substance of linguistics in light of Stalin's articles and the discussion, the Science Section needed more support from scholars in linguistics and in other fields. Kruzhkov and Zhdanov requested that the Secretariat appoint eighteen new instructors, or specialists, to assist them in working out ideological questions in science and to help avoid any mistakes similar to the ones Agitprop had made in linguistics.[73]

Agitprop's call for reinforcements to help with the details of scholarly debates was in part a response to the numerous questions that remained even after the conclusion of the linguistics discussion. Hundreds of letters flooded Agitprop, *Pravda*, *Culture and Life*, and the Central Committee. A select few even earned responses from Stalin, which were published in *Pravda* on August 2, 1950. One letter to Stalin, written by a student from Murmansk, sought to understand what appeared to be a fundamental contradiction between what Stalin had written in the 1930s and what he now espoused. The student wrote: "From your article I understand that out of the hybridization of languages a new language can *never* be formed, while before your article I was firmly convinced that, according to your speech before the XVI Party Congress, under Communism languages fuse into one general [language]."[74]

Stalin defended the apparent contradiction by declaring that the hypothesis that "conclusions or formulas of Marxism, derived as a result of studying one of the periods of historical development, are correct for all periods of development" is "profoundly mistaken." Thus Marx and Engels analyzed nineteenth-century capitalism and determined that the socialist revolution could not be victorious in one country, a conclusion that became a central principle of Marxism. But Lenin, seeing the existence of monopoly capitalism and its weakness, concluded that the socialist revolution "might very well be fully victorious in one country." Stalin saw these contradictory conclusions as applying to two different periods of economic development. Only "Talmudists" would insist on the universal application of laws derived from the analysis of a single system. Stalin offered another example from the classics of Marxism. Engels's formula for the withering away of the state after the victory of the socialist revolution clearly did not jibe with the persistence and strengthening of the state in the Soviet Union. Again, to "Talmudists," who were too formal in their interpretations, there may have seemed to be a contradiction. But to Stalin, Engels was correct for his "own time" and for when socialism would be victorious in a majority of countries. At the XVI Party Congress, Stalin had been concerned with the eventual formation of a single, new language under communism. But in his essay on linguistics, in which he talked about the hybridization of two languages resulting in the triumph of one and the death of the other, he was referring only to the epoch "before the world-wide victory of socialism."[75] Clearly, the implication was that for the time being one language (presumably Russian) would emerge triumphant until a new language could form with the victory of socialism. And maybe not just for the time being: Molotov later recalled that Stalin took up the question of linguistics in part because he believed that after the worldwide victory of communism the Russian language would dominate the globe.[76]

This defense left open the question of which aspects of Marxism were subject to similar contradictions. What in Marxism remained sacred? Stalin suggested that Marxism should not become dogmatic, and his articles on linguistics seemed to lead the way. He concluded by outlining a vibrant Marxism, suggesting that the postwar ideological struggles were about advancing ideology, rather than simply imbibing it:

> Marxism, as a science, cannot stand still; it develops and perfects itself. In the course of its development Marxism cannot help but be enriched by new experience, by new knowledge; consequently, its individual formulas and conclusions must change with the passing of time, must be replaced by new formulas and conclusions corresponding to new historical tasks. Marxism does not recognize immutable conclusions and formulas obligatory for all epochs and periods. Marxism is the enemy of all kinds of dogmatism.[77]

The letters responding to Stalin's articles found in the Party archive attest to the fact that some Soviet citizens took him seriously. One particularly earnest teacher wrote to Stalin:

> You have said more than once that there is dogmatic Marxism and creative Marxism and that you are on the side of the latter. . . . You write that no science can develop

without freedom of criticism and the open struggle of opinions. I am sure that you will allow criticism of your own work. Allow me to expound my view.[78]

He added that he took a Latin saying as his motto: "amicus Plato, sed magis amicus veritas. [Plato is dear, but the truth is dearer.]"[79] Other letter writers took Stalin's call for criticism seriously, even if they did not turn to criticism of the leader's views.[80]

Many of the letters posed questions about the implications of Stalin's work for linguistics, science, and nationality. Even after Stalin's follow-up article, Agitprop identified different sets of questions in the letters that required further explication either by Stalin or by philosophers, linguists, or natural scientists. One group of letters asked questions about language theory, including: What constitutes the form of language and what is its content? How can the dialectical method be applied to the study of language and to what extent is the comparative-historical method compatible with dialectical materialism? Is there any use for an international language, like Esperanto? Is it advisable to create words derived from one or another national language, or should the Russian name-label be used? If the liquidation of capitalism is delayed and communist society is built while the USSR is still surrounded by capitalist countries, will nationalities and languages persist in the USSR or will they solidify into one nation and one common language, even before the victory of socialism in the whole world? Is the following claim correct? "If language is not part of the superstructure, then thought, which is directly connected with language, cannot be considered part of the superstructure. So, thought is not the same as ideology." These questions indicated to leaders at Agitprop that the linguistics discussion had not settled disputes about language in the USSR. Far from it. In fact, it remained unclear who could answer the questions and how.[81]

Other letters ventured beyond linguistics, seeking to understand the implications of Stalin's work for science in general. Was science part of the superstructure or was it, like language, independent of both the base and the superstructure? What was the meaning of Party-mindedness in science? Did the formula of class-based science mean that some truths discovered by science are class-based truths?[82] One letter to Stalin asked for clarification about the relationship between national origins, language, and culture: A. S. Rozenkrants identified himself as the son of Jewish parents who had received Russian education and considered their native language Russian. He considered himself Russian, since his education, culture, and language were all Russian. He worked with Russians. Where was his Jewish national character coming from, if character was based on culture and his culture was Russian? He concluded, "I have Russian, not Jewish national consciousness. My motherland is Socialist Russia." His question to Stalin was shared by many in the Soviet Union at the time: "Who should determine the national affiliation [*prinadlezhnost'*] of a

Soviet citizen whose language, culture, etc. belong to one nation, but whose origin is connected to another? And on what basis should the determination be made?"[83]

Agitprop realized that answering the myriad questions contained in these letters was a daunting task, especially because Stalin himself kept abreast of the work. In August 1950, when Stalin was away from Moscow, he received three different dispatches of letters written in response to his articles on linguistics.[84] Given that Stalin chose not to respond to them, Kruzhkov suggested to Suslov that the best way to deal with the letters was to invite scholars to handle them in their institutes and on the pages of their journals. This seems like a dubious plan at best, considering the volatile nature of the questions asked. And in fact, no answers were forthcoming. There is some evidence that rather than answering the questions, Agitprop tried to limit their pertinence to fields and subjects directly connected with language and culture. On the one-year anniversary of Stalin's articles, the Academy of Sciences planned a general meeting that included the participation of scholars from every section of the academy. Yuri Zhdanov severely curtailed the plans, however, noting that the meeting "would primarily be for show, and would not be productive." Zhdanov nixed the participation of natural and technical scientists. While his reasoning is not known, he seemed to be implying that Stalin's articles did not necessarily inform every field of knowledge. This certainly would have made his job at Agitprop much easier.[85]

While Agitprop could downplay the implications of Stalin's articles for the natural sciences, problems within linguistics still had to be addressed. In the spring of 1951, Agitprop reported to Malenkov that the "restructuring of linguistics research and teaching has not been completed, the dominance of Marr's theory and of his students has not been liquidated. A series of important fields remain in the hands of overt supporters of Marr." Meshchaninov and others had not attempted to correct their mistakes or turn to "practical questions." Likewise, problems persisted at MGU, in the Ministry of Education, and even in Agitprop itself, where one linguistics instructor was evidently not handling his work responsibly. Agitprop suggested a series of meetings to invigorate the field and the creation of a new journal dedicated specifically to linguistics. A few weeks later, the Science Section reported at length to Malenkov, reiterating Agitprop's critical position but couching it within the context of some improvements that had been made. The list of problems in the field was staggering. There were substantive issues: "not a single group of problems of Marxist linguistics has been reworked in light of J. V. Stalin's work"; no work had been done on the development of families of languages; and Russian literary language had not been studied. There were problems with publications—the Academy's literature and language journal published numerous articles containing "serious errors." There were institutional problems—Marr remained influential in Leningrad; republican branches of the Academy of Sciences had not been restructured; and linguistics and Russian-language institutes had not coordinated their work. And there were problems with teaching—new textbooks had yet to be written, and there had been no assistance given to linguistics teachers. A number of courses did not meet scientific standards, and there was little discussion of basic issues in linguistics.[86]

When the Central Committee finally took action, the impetus came from Beria, not Agitprop. By the summer of 1951 Chikobava remained frustrated by the lack of action in the field. Rather than work through Agitprop or the secretaries in charge of science and propaganda—an option he surely could have exercised given the leading role he had played in the *Pravda* discussion— he decided to once again take advantage of his personal connection with Beria. In a letter to his patron he complained about the insufficient restructuring of linguistics in the year since the discussion. Beria forwarded the matter to Malenkov and the Secretariat. In connection with the letter and a general review of the field, a special committee including Vinogradov, Agitprop chief Suslov, his assistant Kruzhkov, and Minister of Education Stoletov recommended the removal of Meshchaninov as head of the Department of Linguistics at Leningrad State University (LGU) and the dismissal of other Marrists from their teaching posts. They also raised the idea of creating a special commission on linguistics that would report directly to the presidium of the academy, thus bypassing the Literature and Language Division. A commission similar to the one proposed for linguistics had been formed in physiology and oversaw that discipline for the remainder of the Stalin period.[87] The linguistics commission would hold regular meetings to evaluate the progress in teaching and research. In February 1952, after looking into the matter, Agitprop decided against supporting such a commission in linguistics, evidently because it would weaken the existing institutional structures in the academy responsible for the field. The publication of *Questions of Linguistics*, beginning in early 1952, was also intended to help solve the problems in the field. The work of linguists in the Union and autonomous republics was of particular concern to Agitprop, as they recognized that there was little coordination between the center and the periphery. The academy formed a special commission to oversee the effort to bring all the linguistics institutions in the country into line with one another. By the middle of 1952—that is, about two years after the linguistics discussion—Agitprop and the Academy of Sciences seemed to have in place the organizational structure to monitor and shape Soviet linguistics.[88]

The question remained, of course, what, exactly, post-Marrist Soviet linguistics was supposed to look like. The simplest thing, it seems, was to criticize the beleaguered Marrists. In 1951 the Academy of Sciences published an anthology of articles under the title "Against Vulgarization and Perversion of Marxism in Linguistics." Rather than mapping out a new direction for Soviet linguistics, the articles set out to discredit the already discredited Marr and his "New Theory of Language." A second volume came out in 1952, and again the central, and almost single, focus was on Marr. (To give a flavor for the volumes, reference to Marr or his work appears in the title of twenty-nine of the thirty-seven articles.) Quoting Stalin's article was obligatory, but making an attempt to further work in the field based on Stalin's suggestions was perhaps too overwhelming a task to undertake except on the most superficial level.[89]

The implications of Stalin's intervention went beyond linguistics, as the initial responses to his articles anticipated. But neither Agitprop nor the academy helped to answer the questions raised by those early responses. In fact, the failure to address the most fundamental questions posed in the wake of Stalin's articles could potentially paralyze more than just linguistics. The editor of the *Great Soviet Encyclopedia* expressed his frustration to Malenkov in diplomatic terms:

> Stalin's brilliant work *Marxism and Questions of Linguistics* gives a deeply scientific treatment of the understanding of the base and superstructure in society, revealing its details and destroying the previous vulgar scheme which placed all spiritual phenomena in the superstructure, and all material ones in the base. In connection with this a question has been discussed for a number of months: what about science? On this question a number of different opinions have been expressed and they often contradict one another. Whatever answer might be placed in the *Great Soviet Encyclopedia* will meet with strong protests from one or another side. . . . We cannot claim in the encyclopedia that the question is being debated, or remains subject to discussion, especially because Comrade Stalin laid out the path for answering it. How should we proceed? How should the question be answered and by whom? Who will determine that the answer is correct and how?[90]

The memo to Malenkov summed up the difficulty of making concrete decisions based on Stalin's articles and highlighted significant tensions in formulating postwar Soviet ideology more generally. Not only were the answers up for debate; the fact that Stalin had broadcast his views made such a debate awkward at best. There was no accepted method of continuing discussion after the time for official discussion had concluded. Stalin's decisive role only deepened the quagmire.

Three overlapping explanations clarify how the Marrists went from nearly complete control of their field in early 1950 to complete defeat a few months later. The first explanation relies in part on personality and patronage. In his articles and speeches from 1946 to 1950 Meshchaninov, the leading Marrist, refrained from forcefully attacking and discrediting non-Marrist Soviet linguists. Though his faith in Marr mirrored Lysenko's faith in Michurin, Meshchaninov never fully adopted Lysenko's winner-take-all approach to academic debate. He appears not to have had the temperament for such the role. Just as important, when it came to connections, it was not Meshchaninov but Chikobava who had a well-situated patron in the Party apparatus. Apparently with Beria's help, Chikobava managed to discuss the situation in linguistics personally with Stalin. Their conversations in April 1950 clearly played a decisive role, first by helping non-Marrists return to previous positions of academic authority, and then in helping crush Marrism altogether. To a great extent, Chikobava's meeting with Stalin explains the timing and prominence of the discussion.

Still, Meshchaninov's relative moderation and Chikobava's access to Stalin can only go so far in explaining the outcome of the linguistics discussion. After all, Stalin had to be amenable to Chikobava's arguments and also had to decide that he would intervene to silence Marrism altogether. To understand the substantive failure of the Marrists, a second explanation is necessary. Marr fell out of favor in 1950 as it became increasingly obvious that his internationalist and economic-based theory, so appropriate for the early 1930s, was completely out of step with the resurgent emphasis on Russian heritage and tradition. Marr's dismissal of pre-Revolutionary Russian linguistics and marginalization of the study of the Russian language could not be reconciled with the broader postwar cultural themes of Russian scientific continuity, Russian chauvinism, and anticosmopolitanism. As Stalin had put it: a formula valid in one epoch may not be valid in the next.

Linguists failed to keep up with ideological shifts in part because they misinterpreted the lessons of the previous scientific discussions. Like many other observers at the time, they failed to notice that Stalin's editorial changes to Lysenko's speech shifted the emphasis from class-based science to "objective" science. As a result, they underestimated the extent to which class categories, such as "bourgeois" and "proletarian," had been displaced by a new hybrid of Russocentrism and objectivity. Having missed the shift from class to nationality, most linguists and even Party leaders assumed that Marr could be made into a direct equivalent of Michurin. But by 1950 Stalin understood that Marr was cosmopolitan and that his disciples had defended untenable and illogical positions simply because they had a monopoly on the title of Marxist linguistics. Thus, the third reason Marrists failed was because Stalin thought they were dogmatic. Ironically, this dogmatism had not existed in linguistics journals or institutes prior to 1947 and 1948. It emerged as a result of genuine efforts by some linguists and Party bureaucrats to respond to previous events in other academic disciplines by defining an ideologically correct doctrine for their field. It must have come as a great shock to them to discover that Stalin now determined that only a field open to "free discussion" could develop in harmony with, and then contribute to, broader trends in Soviet ideology.

Stalin's participation in the linguistics discussion introduced a fundamental paradox that no amount of careful management by Agitprop could make go away. When he called for science to evolve through criticism and the free exchange of opinions, Stalin did so with the presumption that scientific truth would mesh with Marxist-Leninist doctrine. But the ultimate, and in the end only, confident interpreter of the doctrine was Stalin himself. Criticism and the free exchange of opinion could not produce a truth more powerful than the ones declared by Stalin. Thus his statements on linguistics became both the starting points for further research and the only safe end points linguists could reach in their conclusions. Instead of encouraging the free exchange of scholarly opinions, his articles had the opposite effect, prescribing the number of legitimate topics in the field. Concentrating scholarly authority and Party authority in one body—the "coryphaeus of science"—did not solve the tension between scientific and political truth any more than attempts that had been made in previous discussions. Instead, it amplified the irony of Stalin's dictating answers in the name of the free exchange of opinions.

CHAPTER 6

"ATTACK THE DETRACTORS WITH CERTAINTY OF TOTAL SUCCESS"

The Pavlov Session of 1950

At the same time Stalin's articles on linguistics abruptly silenced the movement to present Nikolai Marr as a paragon of Soviet scholars, the coryphaeus supported a parallel effort to unify physiology behind the work of the Soviet Union's most famous scientist, Ivan Pavlov. This time the climactic event was the joint meeting of the Academy of Sciences and the Academy of Medical Sciences in June and early July 1950. Hundreds of Soviet physiologists attended; the whole country followed its progress. *Pravda* replaced its weekly coverage of the linguistics discussion with daily reports from the Pavlov Session—as the meeting became known. In this session, unlike the previous scientific discussions, Stalin and his assistants did not merely adjudicate the disputes brought forth by scholars. Instead, Yuri Zhdanov, the Party's lead man in science, took the initiative. A few years earlier he had brazenly challenged Lysenko's effort to divide biology without appreciating the extent to which doing so brought him into direct conflict with Stalin. Now, relying on the coryphaeus' approval and advice, he identified Pavlovian science as the only true Soviet physiology and confidently repressed any alternative interpretations.

Previous discussions had taught Yuri Zhdanov that scientific disputes overlapped with debates about doctrinal orthodoxy, concerns about loyalty to the Soviet state—often measured in terms of professed patriotism and distance from Western ideas and scientists—and the emphasis on applying science to practical problems. In these circumstances, Pavlov proved to be a stable model

of a Soviet scientific hero. In 1904, he had become Russia's first Nobel laureate, yet unlike many prominent Russians he did not leave after the Bolsheviks
took power in 1917.[1] Indeed, Lenin personally supported him and his research
even during the harsh years of the Russian Civil War, in part because of the
prestige the great scientist brought to the regime. As the historian Daniel
Todes suggested, in this way both Pavlov and the Bolsheviks managed to get
what they wanted from one another.[2]

In the postwar period, Yuri Zhdanov recognized that Pavlov's worldview,
although not explicitly Marxist, could be used to bolster the main ideological
tenets of Soviet science. Pavlov had insisted that his approach to physiology
was materialist (for him this meant disdain for the subjective methods of psychologists) and emphasized the importance of applying physiology to medicine.[3] The ideological confluence ran deeper. Pavlov's research on conditioned
reflexes, which he claimed provided the basis for analyzing even the most
complicated forms of human and animal behavior, corresponded to the Party's
interest in the transformation of Soviet man.[4] Just as Pavlov argued that stimuli and conditions were at the root of animal behavior, controlling the stimuli
and conditions of Soviet society could help revolutionize Soviet citizens'
moral and material worlds.

In the 1920s and 1930s Pavlov's ideas had grown increasingly isolated from
physiology as it was developing in the West. This only strengthened his appeal
to Yuri Zhdanov. In the 1940s, the Party asserted the independence of Soviet
science from capitalist science. It certainly helped that, during the last years
of his life, Pavlov reconciled some of his obvious differences with the regime
and willingly associated with those Soviet physiologists who endorsed dialectical materialism as the only sound philosophical basis for science. He also
publicly acknowledged the value of the Soviet experiment. Enough elements
existed for Party propagandists to assert that Pavlov's methodology and results,
like Lysenko's, were uniquely Soviet and would therefore serve as a model of
progressive science for the world. This was the sort of campaign Zhdanov had
in mind.

Zhdanov faced a major obstacle, however. Propaganda aside, Soviet physiologists, like those in the rest of the world, had become increasingly critical of
some of Pavlov's ideas. As early as the 1920s, scientists challenged the primacy
of conditioned reflexes for explaining behavior. The Georgian physiologist
Ivan Solomonovich Beritashvili (also known as Beritov, the Russianized version of his name) believed that Pavlov's interpretation of neural processes was
overly simplified. Before 1917, Beritashvili worked as a disciple of Pavlov's in
St. Petersburg and also spent a year in Holland studying with the neurophysi-
ologist Rudolph Magnus. In 1919 he returned to Georgia, where he founded
a physiology laboratory at Tbilisi University. There his experiments challenged both Pavlov's typology of reflexes and his ideas about their physiological transmission. Pavlov disdained the notion of instincts and had claimed

there were simply two types of reflexes, unconditioned and conditioned. He discovered that dogs could be trained to salivate whether the stimuli were benign, like bells, or painful, like electric shocks. This proved, he said, the similarity of all conditioned reflexes. In his terminology, stimuli were completely "indifferent." But Beritashvili's experiments indicated that dogs were more easily conditioned to respond to certain biologically significant stimuli, such as the appearance of food, raising questions about the binary classification Pavlov used for all reactions.[5]

Furthermore, some of Pavlov's theories were not sufficiently sophisticated to explain other phenomena Beritashvili discovered in his lab. For instance, Pavlov had postulated that stimuli traveled along affector neurons to the cerebral cortex, at which point another stimulus would travel along effector neurons signaling a reaction. Repeatedly presenting the sound of a bell with the appearance of food would create temporary connections in the dog's nervous system, until the dog would salivate upon hearing the bell regardless of whether food was presented. If not reinforced, or if the bell was repeatedly rung without food, the dog would slowly loose this temporary reflex. Pavlov called this the "reflex arc," from stimuli to cerebral cortex to reaction. He declared that the reflex arc formed the fundamental process behind all animal behaviors. Using a set of experiments, Beritashvili challenged Pavlov's theory. For instance, a mouse learned to run through a maze to find food, thus presumably reinforcing the temporary connections between the cerebral cortex and the legs. Yet if the mouse's legs were paralyzed, it would still somersault toward the food, thereby suggesting that the process of conditioning did not correspond to Pavlov's concept of the reflex arc. Other experiments showed that dogs whose cerebral cortexes had been removed, and indeed animals without brains or even spinal cords, all adapted to stimuli. If conditioned reflexes could be shown to develop outside the areas of the brain responsible for higher functions, it was difficult to see how they could be shown to form the bases of all mental activity.[6]

The Soviet scientific community widely recognized the importance of Beritashvili's work. He was awarded the Pavlov Prize in 1938, was elected to full membership in the Academy of Sciences in 1939, and received a Stalin Prize in 1941.[7] Like Beritashvili, Lina Solomonova Shtern, the Soviet Union's most famous woman scientist, challenged some of Pavlov's ideas in the 1920s and 1930s. Shtern, a Jew born in 1878 in what is now Latvia, studied at the University of Geneva, where in 1917 she became a professor of physiology. Acting on her political sympathies, in 1925 she moved to the Soviet Union and became the director of the Institute of Physiology in Moscow. In contrast to Pavlov's emphasis on the nervous system, Shtern analyzed the chemical basis of physiological processes such as digestion and respiration in animals and humans. She also received accolades: she joined the Party in 1938, became the first female member of the Academy of Sciences in 1939, won a Stalin Prize in 1943, and joined the Academy of Medical Sciences in 1944.[8]

Beritashvili's and Shtern's professional successes suggested that up to and through the Second World War Soviet physiology could not simply be equated with Pavlov and his school. Part of the reason was that Leon Abgarovich Orbeli, Pavlov's leading disciple, had adopted an inclusive understanding of his mentor's legacy. Orbeli, an Armenian born in 1882, worked closely with Pavlov during the first decade of the twentieth century, when they conducted research on conditioned reflexes. On Pavlov's recommendation, he also spent two years before the Revolution working in physiology laboratories in Germany and England. After Pavlov's death, in 1936, Orbeli assumed many of his mentor's administrative positions, a clear indication that he had earned the Party's trust.[9] He was a member of the Academy of Medical Sciences and the Military Medical Academy, and after 1940 he headed the Biology Division of the Academy of Sciences. Beritashvili and Shtern clearly benefited from Orbeli's stewardship of the field and his acceptance of a variety of approaches to physiology.

Even within the Pavlov circle, interpretations of the master's work varied. Another of Pavlov's disciples, Petr Kuzmich Anokhin, openly agreed with much of the criticism coming from physiologists outside Pavlov's immediate sphere of influence and turned his attention to the relationship between the central and peripheral nervous systems. Others who had worked closely with Pavlov, however, such as Konstantin Mikhaelevich Bykov and Anatolii Georgievich Ivanov-Smolenskii, wanted to maintain a stricter adherence to their teacher's work. Bykov was only four years younger than Orbeli but, after Pavlov's death, he failed to gain much control over the field. Orbeli even objected to his election to the Academy of Sciences in 1946.[10] Ivanov-Smolenskii, a doctor in the Red Army during the Russian Civil War, joined Pavlov's lab in 1921, when the great physiologist's best work was well behind him. Even after Pavlov's death, Ivanov-Smolenskii, unlike Orbli, continued to work solely on research projects outlined by Pavlov himself.

This diversity of opinions about and approaches to physiology, however, was hidden behind the veil of obligatory public praise for Pavlov and his work. The historian David Joravsky summed up the situation before the war: "The ideological establishment did not pry into the work of Pavlov's disciples, to see if they were true to the grand old doctrine, and they did not attempt sophisticated correction to the popular notion that Pavlov's doctrine was the final word in brain science."[11] After the Second World War, the Party took increasing interest in reconciling science with public rhetoric. This had spurred the effort to bring linguists in line with Marr's teaching. In physiology this meant ironing out the differences among Pavlov's disciples as well as challenging the views of Beritashvili and Shtern. Orbeli and other physiologists may have been reluctant to force a rigid interpretation of Pavlov's work onto their colleagues, but the Party, and especially Yuri Zhdanov, had no such qualms. As a Party bureaucrat he could not lead the charge openly. Instead, he needed to find a candidate among Soviet physiologists who would be willing to take on the establishment, as Lysenko had done in biology.

Orbeli, the USSR's most prominent physiologist after Pavlov's death, was reluctant to suppress new ideas in his field. As Soviet propagandists divided the world into two camps and Russian chauvinism grew stronger in the USSR, Orbeli's position grew weaker. In 1947, he headed the Soviet delegation to the International Physiological Congress meeting in London. When news reached the Central Committee that Orbeli and others had delivered their speeches in English rather than Russian, Party secretary A. A. Kuznetsov was furious.[12] Orbeli's standing with the Party further slipped in 1948 when he failed to support Lysenko or research into the inheritance of acquired characteristics.

On the eve of the August Agricultural Academy session, Yuri Zhdanov published a veiled attack on Orbeli in an article in *Culture and Life*, the organ of Agitprop. He complained that physiologists wasted time publishing works by Beritashvili and Shtern, rather than propagating the "ideas of the great Russian physiologist I. P. Pavlov." During the meetings at the Academy of Sciences and the Academy of Medical Sciences organized in response to the revelations at the Agricultural Academy session, Orbeli was singled out. Among other serious errors, he had created a friendly setting for "formal genetics," and he had undervalued Pavlov's research on the inheritance of conditioned reflexes.[13] This last point was crucial. Lysenko and his supporters argued that Pavlov, like Michurin, had rejected formal genetics. Their evidence was based in part on one of Pavlov's student's attempts before the Revolution to show the inheritance of conditioned reflexes in mice.[14] But, unlike Orbeli, they blatantly ignored Pavlov's own efforts to distance himself from this work and its implications.[15]

Orbeli lost his post as head of the Academy of Sciences' Biology Division in the restructuring that occurred in the aftermath of the Agricultural Academy session.[16] Still, he avoided a total rout. He maintained a number of posts that would allow him to defend physiology against a Lysenko-like campaign. And out of political expediency he included the study of the inheritance of conditioned reflexes in research plans.[17]

Efforts at reconciliation did not deter the Party machine. As the political demands on Soviet society became increasingly rigid in the late 1940s, physiologists began to suffer. Newspapers accused Shtern, along with many other prominent Jews, of conspiring to create a Jewish state in the Crimea. The Security Ministry arrested her in January 1949. Her science was quickly dismissed as cosmopolitan; her institute was merged with the Bekhterev Institute to form a new Institute of Physiology of the Central Nervous System. Bykov became the new director. His strict adherence to Pavlov's agenda and refusal to ridicule Lysenko began to pay dividends. Stalin greeted the restructuring of the institutes with excitement, noting, "We are guilty of confining Bykov's luminous science to the shadows, while a random meteor like Shtern can occupy leadership positions. It is time to be done with such abomination."[18] When Orbeli and a group of scientists from the Bekhterev Institute wrote to Stalin in protest, he paid them no heed. Bykov, the new Party favorite, refused to allow the scientists recently placed under his charge to continue their old research agendas. Yuri Zhdanov and Agitprop supported his decision and concluded that Bykov would help lead Soviet physiology back to Pavlov's ideas.[19]

Shtern's arrest and the dissolution of the Bekhterev Institute did not bode well for those who hoped to maintain a broad view of physiology. But it was not clear how the Party would go about diminishing Orbeli's obvious power in the field. As late as June 1949 Orbeli became a member of the Academy of Sciences' presidium over the objections of Topchiev, the Party's main representative at the academy. Sergei Vavilov, the academy's president, clearly supported Orbeli, whom he considered an authoritative and valuable scientist.[20] As a presidium member, Orbeli maintained contact with some of the USSR's most powerful academic administrators.

Vavilov also reported in a memo to Stalin and Malenkov that Orbeli had earned a strong reputation among his fellow scientists and that his Pavlov Institute was conducting important work. Even Bykov and Oparin (the Lysenkoist who had replaced Orbeli at the academy's Biology Division) agreed. While Orbeli may have spread himself too thin by trying to hold too many administrative positions, Vavilov insisted that "in the opinion of our physiologists and biologists in general, Academician Orbeli . . . is at the present time the strongest representative of physiology in our country."[21]

In February 1950, the Politburo approved the Ministry of Defense's proposal to remove Orbeli as the head of the Military Medical Academy. Orbeli's "personal request in connection with [his being] overburdened with other work" was the official reason for the decision.[22] With Orbeli under pressure, this seemed to be a compromise position. He would give up some of his posts but maintain his reputation among scientists. Bykov's obvious reluctance to attack Orbeli with the anti-Soviet, antimaterialist epithets used so frequently in the biology discussion suggested that he would not be the best candidate to lead an anti-Orbeli campaign anyway. But Zhdanov was not easily deterred. Upon receiving a copy of Vavilov's memo, he suggested to Malenkov that the Central Committee consider a major overhaul in Soviet physiology just the same.[23]

While restructuring the field required considerable bureaucratic planning and finesse, a campaign to promote the myth of Pavlov proved fairly straightforward. The primary goal was to present Pavlov as the world's foremost physiolo-

Figure 10. A 1949 Soviet postage stamp celebrates the "100th birthday of the great Russian scientist academic I. P. Pavlov."

gist. His science provided the blueprint for future research in the USSR and the rallying point for combating bourgeois and Western approaches in the field. As Orbeli, Vavilov, and Zhdanov maneuvered behind the scenes, the public campaign expanded with both a feature biographical film and, in September 1949, a nationwide celebration of the hundredth anniversary of Pavlov's birth. The repercussions for Pavlov's disciples were potentially grave. Pavlov was portrayed in such glowing terms that it was unlikely any living Soviet physiologist could compare.

In early 1949, the Ministry of Cinematography proudly reported to Stalin and Agitprop that the new film *Academician Ivan Pavlov* was ready for distribution. The ministry's artistic council had carefully edited the script, watched the film many times, and had even taken into consideration critical comments from scientists. In the council's opinion, the film showed "Pavlov's struggle with reactionary trends in physiology and his hatred for idealist pseudoscience."[24]

In the press, critics praised the film for showing the Russian physiologist's patriotism. The actor playing Pavlov stated in the film, "Science has a fatherland, and a scientist is required to have one too! I, dear sir, am Russian and my fatherland is here!" The film's Pavlov believed that science must be geared toward practical results, and he recognized the authority of the Party,

Figure 11. The physiologist Leon Orbeli giving a speech at the celebration of Pavlov's 100th birthday in September 1949. Courtesy of RGAKFD.

which created the best conditions for science in the world. The film won a top Stalin Prize.[25]

The emphasis on Pavlov as a Russian, as opposed to Soviet, scientist is exemplary of the persistent Russian chauvinism of the late Stalin period. Beginning in the mid-1930s, Soviet ideologues were increasingly comfortable blurring the Russian imperial heritage with Soviet patriotism. The same nationalist impetus that saw newfound enthusiasm for Ivan the Terrible and Pushkin, as well as Mendeleev and Lobachevskii, made it possible to portray Pavlov as a Russian hero without that having negative connotations. As a result, however, Soviet physiology splintered further: Bykov was Russian; Orbeli, Shtern, and Beritashvili were not.

The recasting of Pavlov's legacy continued. A few months after the film's release, the Central Committee put secretary Mikhail Suslov, one of the

USSR's leading authorities on ideological issues, in charge of organizing the celebration of Pavlov's hundredth birthday.[26] Suslov's involvement and the scale of the celebration show that Pavlov's image was of national importance. Plans called for meetings in Moscow, Leningrad, and Pavlov's hometown of Riazan; scientific meetings at the republican academies of sciences; the republication of Pavlov's work; the creation of a Pavlov Medal in physiology; and a documentary film on Pavlov's life. The original draft of the Politburo resolution outlining the celebration even called for renaming the 850-year-old city of Riazan after Pavlov. Showing some restraint, the final resolution did not accept this last suggestion.[27]

Agitprop carefully monitored the plans for celebrating Pavlov's birthday— and found little room for Orbeli. But other physiologists did not immediately strike the combative tone Agitprop was after. V. S. Kruzhkov and Zhdanov read the speeches that Bykov and others planned to give at the Bolshoi Theatre in Moscow and the Kirov Theatre in Leningrad. They reported to Malenkov that Bykov's speech was strong, particularly because it emphasized the "influence of the ideas of the revolutionary democrats on the consciousness of Pavlov and his Russian teachers." Of course, this also allowed Bykov to imply that Pavlov himself was sympathetic to revolutionary activity. As for the Soviet period, Bykov praised the government's support of Pavlov's work after the October Revolution, while also mentioning Pavlov's materialism and struggles with idealism in biology. According to Zhdanov and Kruzhkov, the only problem was that Bykov had not attacked wayward disciples, such as Orbeli, or competing approaches, such as Beritashvili's.[28] Other proposed speeches presented similar problems.[29] Again, physiologists preferred to present their field as united behind Pavlov's teachings; Party leaders, in contrast, wanted the celebration to be more divisive.

On Pavlov's birthday a front-page editorial in *Pravda*, "A Great Son of the Russian People," instructed readers that Pavlov was "close to the heart of every Soviet person" and had been personally supported by Lenin and Stalin. He had founded a "new epoch in physiology" and continually struggled against "reactionary, idealist, false theories . . . of the bourgeoisie in the United States, England and other capitalist countries." Pavlov loved science, but, like Michurin, he never forgot that science was first and foremost a means for deciding practical problems and controlling nature. In a particularly creative reading of the history, the editorial emphasized that Pavlov's work "expanded especially quickly after the Revolution" and that he "selflessly served his socialist motherland and his people."[30] The more the propaganda campaign praised Pavlov, the worse the current state of Soviet physiology looked in comparison.

For Zhdanov the film and the birthday celebration were only preparatory steps for restructuring Soviet physiology. On September 27, Pavlov's hundredth birthday, Zhdanov wrote to Stalin to complain that despite the celebration no one was bothering to ask the question of whether Pavlov's science was being correctly developed in the USSR. Clearly he thought it was not, and he asked Stalin to read an eighteen-page report he had written on the situation. He also sent a copy to Malenkov.[31]

Zhdanov reported significant problems in Soviet physiology, which, almost by definition, came from those who were opposed to Pavlov: "Anti-Pavlovian tendencies have strengthened considerably in Soviet scientific literature over the last few years. Revisionist trends, crude misrepresentations of Pavlov's work, efforts to show that this work has become outdated and is meaningless for the development of physiology, psychology and medicine have all spread."[32] Zhdanov saw this problem stemming in part from those who were explicitly anti-Pavlovian, such as Beritashvili and Shtern, and in part from a general the lack of emphasis on the application of Pavlov's work to medicine. Beritashvili's "anti-Pavlovian revisionism" aided "bourgeois science in the West" and its "furious attacks on Pavlovian science as a materialist science."[33] Shtern and other "rude and vulgar mechanistic" physiologists needed to be "unmasked" and "exposed" for their "anti-Pavlovian orientation."[34] Zhdanov was also concerned that medical institutes did not teach Pavlov's ideas and that clinicians were unfamiliar with the application of Pavlov's work to medicine. But the most important problem in the field had to do with Orbeli's "monopoly."[35]

Zhdanov reported to Stalin that Orbeli was under the influence of foreign physiologists and that he had committed "serious anti-Pavlovian and anti-Michurinist mistakes in the study of inheritance." Zhdanov also held Orbeli responsible for supporting Beritashvili and Shtern, while using his "monopoly of the leading physiological institutes" to do everything in his power to prevent the advancement of Pavlov's disciples. Evidently Orbeli had tried to block other scientists' advancement in the Academy of Sciences; Bykov became a member despite Orbeli's efforts. Zhdanov suggested, "It is necessary to liquidate Orbeli's monopoly of the development of Pavlov's science and to subject his mistakes to criticism."[36]

In concluding his memo, Zhdanov outlined what he believed needed to be done to improve Soviet physiology. He called for the restructuring of the higher-education curriculum to emphasize the importance of Pavlov's work for medical research and practice, particularly in psychoneurological clinics. But, most important, he called for a conference where Bykov would give a speech criticizing Orbeli and others while defining the parameters of Pavlov's work. Significantly, even at this stage Zhdanov had in mind precisely the kind of negative comments that had been relatively absent from the birthday celebration. The purpose of the meeting was to "subject efforts at revising or changing Pavlov's work to deep criticism . . . [and] to isolate Pavlov's enemies."[37]

Stalin clearly endorsed the memo's analysis of the situation and was pleased that Zhdanov had "addressed the Pavlov issue." He wrote to Zhdanov on

October 6, 1949, "I do not have a single disagreement with you on any of the points you raised in your letter." If anything, Stalin was even more critical of Orbeli: "In my opinion Academician Orbeli has caused the greatest damage to Pavlov's science. By falsely appointing himself as Pavlov's most important student, Orbeli did everything he possibly could to disgracefully silence Pavlov with provisos and ambiguities. [Orbeli's] cowardly and disguised raids against Pavlov [constituted an effort] to dethrone and slander him." According to Stalin, Beritashvili and Shtern were not as dangerous, because they attacked Pavlov openly. Stalin agreed with Zhdanov about what should be done: "The sooner Orbeli is denounced and the more soundly his monopoly is liquidated, the better."[38]

Ever the political strategist, Stalin tutored Zhdanov on how he should go about organizing Orbeli's demise. He agreed that Bykov was the best candidate for the role of Pavlov's leading disciple. But was he combative enough to help purge the field of Orbeli and his followers? "It's true," Stalin admitted, "he is a little timid and doesn't like to mix it up." Rather than find a more steadfast candidate to depose Orbeli, however, Stalin relied on Zhdanov to toughen Bykov up. "Support him, and if he isn't manly enough, fix things in such a way that he'll join the battle. Explain to him that without a melee it will be impossible to defend Pavlov's great work." Stalin also offered "a word or two on the tactics" to be used in the upcoming fight. "At first you need to surreptitiously gather Pavlov's supporters, organize them, distribute roles and only after that can you gather the actual meeting of physiologists that you're talking about. That's where you can bring the broad battle to [Pavlov's] detractors. Without this, it is possible to mess the whole thing up. Remember: attack the detractors with certainty of total success." Ensuring that Zhdanov would not be isolated in his effort, Stalin forwarded Zhdanov's original memo and his own response to Malenkov along with a note stating, "I think that the Central Committee should fully support this."[39]

Following Stalin's advice, Zhdanov spent the end of 1949 and the beginning of 1950 gathering material for the conference. Bykov went to work drafting his speech but, even with Zhdanov's editorial directions, he remained reluctant to attack Orbeli head-on. Meanwhile, Zhdanov solicited speeches from a group of "strong Pavlovian physiologists" who joined the criticism of Orbeli and the state of Soviet physiology more generally. He also made sure he had the support of the minister of health. By early April, Zhdanov was pleased enough with the progress to take the next step. On April 13, he forwarded the latest draft of Bykov's "Developing Pavlov's Ideas (Tasks and Perspectives)" to Stalin. As he explained to his boss, Bykov's speech still did not consistently attack Orbeli. But he assured him that these and other errors "can be easily eliminated." In keeping with the strategy they had worked out in October 1949, Zhdanov reported to Stalin: "I believe that enough power has now been amassed to turn to an open, organized offensive against the enemies and hypocritical 'friends' of Pavlovian science."[40]

Stalin agreed, and Zhdanov began to set the offensive in motion. On April 25, he sent Malenkov a detailed report proposing an early June meeting to take place under the auspices of the Academy of Sciences Biology Division and the Academy of Medical Sciences. He planned to invite the members of both academies as well as five to ten people from each of the major physiology institutes in the country. When scientists from leading medical and psychiatry institutes and disciples of Pavlov working in the periphery were included, the total number of participants would be about four hundred. The venue would be the House of Scientists, the site of the Agricultural Academy session two years before. In addition to an introductory statement by Sergei Vavilov and Bykov's report, Zhdanov called for the meeting to begin with four scientists' reports that he already had on file in the Science Section. To show how thoroughly the offensive had been organized, Zhdanov also mentioned eleven other "defenders of the correct views" who were prepared to speak.[41]

Ideally, the targets of ridicule would also attend. Zhdanov wrote, "It makes sense to insist that the scientists subject to criticism should also speak. Here I mean first of all Orbeli and his allies." Clearly, scientific discussion required debate and "free and open criticism," as Stalin had put it, even if political authorities had predetermined the outcome. Zhdanov also envisioned a press campaign prior to the session, with selected parts of Pavlov's letters and speeches reissued in *Pravda*, *Culture and Life*, and elsewhere. Finally, Zhdanov put together an organizing committee (with himself as a key member) to take care of the details.[42] The memo suggests that, following Stalin's advice, Zhdanov had coordinated the attack on "Pavlov's detractors" to maximize its chances of success.

By the end of May, the Politburo had approved a revised proposal removing two of the plenary reports. The Politburo evidently concluded that reports by Bykov and Ivanov-Smolenskii would suffice to get the session under way. They also increased the scope of the press campaign, adding to the number of reprints of Pavlov's articles and expanding the number of journals that would print them. The meeting's time was pushed back to late June and early July. The Politburo circulated its decision about the meeting to all the members of the Orgburo, the editors of *Pravda*, *Culture and Life*, and the *Medical Worker*, and the heads of the two academies.[43] Minor changes aside, the organizational structure and content of the meeting remained consistent with Zhdanov's initial report and with the tactical advice Stalin had given him the previous fall.

The Science Section's goals for the Pavlov Session were much more consistent than they had been in the other discussions. Zhdanov, and to a lesser extent his immediate superiors Kruzhkov and Suslov, kept Stalin informed of their plans for physiology, and this no doubt left little room for surprises.[44] In this case, Stalin did not meet personally with scientists or assert any scholarly expertise of his own on the subject. But he certainly affected the meeting in other ways. Not only had Stalin read Zhdanov's initial report and offered tactical advice about how to go about attacking Orbeli and others; he also read and edited Bykov's speech.[45] Stalin even kept abreast of minor details concerning the attack. On the eve of the meeting, he and the Politburo approved Agitprop's request to increase the number of days *Pravda* would cover the event.[46]

Although archival materials on the organizing committee of the Pavlov Session are limited (especially when compared with the complete records on the planning of the 1949 All-Union Conference of Physicists), the minutes of the meetings that are available display the extent Zhdanov and others were able to script the session.[47] The Academy of Sciences was represented by its president, Vavilov, and its academic secretary, Topchiev. The Academy of Medical Sciences was represented by its president, Anichkov, and vice president, Razenkov. Zhdanov also attended the organizational committee meetings, though evidently he rarely spoke.[48] Despite the power of the committee members, they dealt mainly with mundane issues.

The minutes also make clear that despite Zhdanov's months of preparation, many aspects of the meeting remained unplanned. On the eve of the session, Vavilov was still fretting that because the outcome was unknown—at least to the vast majority of participants—it would be difficult for participants to strike the correct tone in their speeches. Logistical problems also abounded. The Central Committee had suggested that around 400 people participate, with hundreds of others filling up the House of Scientists. But the organizing committee invited 485 people and was having a hard time keeping the number of actual participants to 450. They were evidently receiving upward of one hundred telegrams a day from people asking to be invited. Others simply showed up in Moscow from out of town demanding tickets and hotel rooms.[49]

The organizing committee was not confident that the so-called anti-Pavlovians would participate. As Topchiev noted, "Some people absolutely need to be there." If they declined, then either Vavilov or Anichkov would have to insist on their presence. The organizers were particularly concerned about Beritashvili. Topchiev reported to the committee: "I specifically asked N. I. Muskhelishvili [president of the Georgian Academy of Sciences] about Beritashvili: would he be there? N. I. Muskhelishvili said, 'I doubt it.' I said, measures must be taken so that he will be there. N. I. Muskhelishvili said that he would call. We need these kind of people to be there."[50] Considering that Beritashvili was one of the main representatives of the anti-Pavlovian trend in Soviet physiology, and therefore one of the main targets of the session, his absence could certainly change its tone. In order for the session to maintain the guise of a scientific discussion, both sides would have to participate.

A certain lack of coordination is to be expected in organizing a conference with over a thousand participants. Given the number of people who wanted to speak, and Vavilov's reluctance to "limit people's freedom," the organizing committee realized that the meeting would last one day longer than originally planned.[51] They also had hoped to distribute Bykov's keynote speech to all the participants in advance of the session, but printing delays prevented them from doing so. An exhibit on Pavlov's life and work, which was supposed to be reviewed by the organizing committee the day before the session, was not prepared in time.[52]

The organizing committee wanted to create the right conditions for the meeting. In addition to busts of Lenin and Stalin, Pavlov's portrait was displayed in the main hall "in order to immediately give the participants the idea that the whole conference will be conducted under the sign of Pavlovian science." Exhibits outside the meeting hall would emphasize Pavlov and not anyone else.[53] The point was to create a "scholarly atmosphere." Vavilov rejected a proposal to film the session, stating, "In my opinion, there is absolutely no reason for it. These are not actors, it will only interfere with things. . . . This is a businesslike session, it is not necessary to weigh it down with that."[54]

Heated disagreements concerned the nature of the meeting's stenographic record, which would provide the official version of what was said at the session One member of the committee proposed that the minutes be read and edited by a special committee while the session was taking place. But Vavilov was worried that such a quick turnaround might leave some people vulnerable to attack for things said during the heat of the moment. Instead, a carefully edited record, prepared after the general themes of the meeting had been clarified, would be more advantageous. When told that the committee would only begin the process of editing, Vavilov responded: "How can you edit from the beginning when you don't know the ending of the session? That's impossible! . . . Don't forget this is a session for debate. The most polemical, shrill things might be said by one or the other side. Can we really [immediately edit the stenographic record]?!" The committee agreed that publishing a "raw stenographic record" was not advisable, and a committee was formed to help with the editing.[55] Vavilov's concern suggests that some of the content of the meeting remained up for grabs. If certain scholars did not play their roles, Vavilov wanted the organizers to have the chance to make the meeting look more successful, and he also wanted to protect those who might become inadvisably shrill.

Historians have often compared the Pavlov Session to the August Agricultural Academy session, but the other scientific discussions of the postwar period must also have had an impact.[56] Topchiev, after all, had chaired dozens of meetings in anticipation of the physics conference in 1949. He and the organizers had spent months planning each detail of that conference, reviewing each speech line by line, only to have the whole thing canceled. In comparison, the plans for the Pavlov Session must have seemed rushed and relatively open ended.

In its scale and the attention given to it in the USSR's leading newspaper, the Pavlov Session closely resembled the Agricultural Academy Session of two summers before. The tone and substance of the Pavlov discussion, however, suggested that this was a different sort of meeting. First, the plenary reports did not come close to matching Lysenko's vitriol. Second, the critics and their targets (with the possible exception of Beritashvili) agreed in principle that Pavlov's teachings provided the beacon for all further work in physiology. Although speakers accused Orbeli and others of idealism and dangerous acceptance of Western, bourgeois concepts, there was no physiological equivalent to "Mendelism-Morganism." Indeed, the physiology session more closely resembled the philosophy discussion, in which the reigning head of the discipline, Aleksandrov, stood accused of not living up to the standards of Marxist-Leninist philosophy. Now it was Orbeli's turn to recognize that he had gone astray, leading physiology down a path that diverged from the "immense program" mapped out by Pavlov.

Significantly, the organizing committee met for the last time only days after Stalin's first article on linguistics appeared in *Pravda*. Vavilov, at this point experienced in the political intrigue surrounding postwar academic disputes, must have been acutely aware of how quickly the tone and content of a scientific discussion could change. In 1946, Aleksandrov had been praised as the nation's leading philosopher and won a Stalin Prize; by 1947 he was the central figure in a controversy enveloping all of Soviet philosophy. Likewise, one minute the Academy of Sciences was expected to support Marr; the next it was supposed to criticize him. No doubt it looked as though Bykov had received official support from the Party and that Orbeli was on the way out, but could Vavilov be confident of that outcome before the session had been played out? Given the surprise associated with the linguistics discussion and Stalin's direct participation, it is likely that some members of the organizing committee and even many of the participants in the discussion took the stage under a similar shadow of apprehension and uncertainty.

The joint Academy of Sciences and Academy of Medical Sciences "Scientific Session on the Physiological Teachings of Academician I. P. Pavlov" began at the Moscow House of Scientists on June 28, 1950. There were more than one thousand participants and guests in attendance from over fifty cities and all the republican academies of sciences. The overall cost of the session came to nearly a half a million rubles, including train fares, hotel rooms (guests stayed at the elite hotels Moskva, Grand, and Evropa), buses, stenographers' fees, flowers, and posters. *Pravda* began daily coverage of the physiology session on June 29 and continued publishing speeches from the meeting through July 9.[57]

Sergei Vavilov's introductory remarks clarified that he hoped the meeting would concentrate on scientific issues. In the broadest brushstrokes, he set the agenda: this meeting, in contrast with events marking Pavlov's birthday the year before, was not for "celebrations, historical observations, and reminiscences, but for a critical and self-critical examination" of how matters stood "with regard to the development of Pavlov's legacy in the Soviet Union." To his mind, this was a scientific meeting. He noted that though the Soviet government had "created unprecedented conditions for the advancement" of Pavlov's work, his heirs had not followed Pavlov's lead. For instance, very little effort had been made to develop Pavlov's concept of the "second signal system," which supposedly explained the development of speech and other complex brain activity in humans. Referring directly to Stalin's article on linguistics, Vavilov called for participants in the session to engage in a "creative clash of opinions and free criticism, without regard for established authorities [and] undeterred by long-standing traditions." Still, his remarks avoided attacks against specific scientists.[58]

Razenkov, the vice president of the Academy of Medical Sciences, spoke next. Unlike Vavilov, he saw the meeting as an opportunity to expose the political and ideological mistakes of specific physiologists. He pointed out that Lenin and Stalin had supported Michurin and Pavlov and that with the help of Lysenko, Michurin's work had prevailed over "reactionary, idealist trends in biology." After a brief bit of self-criticism, Razenkov made clear that the real enemies were those who supported anti-Pavlovian concepts, either by citing Western works or by openly questioning the validity of some of Pavlov's ideas: Beritashvili was "notorious as an opponent of Pavlov"; Orbeli had not "developed Pavlov's work" and allowed Morganist-Weismannists to "ply their trade" at the institute. Likewise, Anokhin "has had an infatuation with fashionable reactionary theories." Bykov, on the other hand, was "doing much to develop Pavlov's theories." His few mistakes were minor. Despite the thrust of his speech, Razenkov did not call for an all-out attack on Orbeli and others: "Our constructive and comradely criticism should not have anything in common with the vicious, hostile criticism that our Party has always condemned."[59] In other words, accusations of political and ideological mistakes had to be couched in scientific and objective terms. With the audience warmed up, Bykov, the new heir to Pavlov's throne, took the stage.

The plenary reports, by Bykov and Ivanov-Smolenskii, set the parameters for the discussion, as Zhdanov had planned. Bykov went first and began with references to the "victory of Michurinism" and Pavlov's role in striking a "blow to idealist physiology." He hardly mentioned Beritashvili or Shtern. Orbeli was his main target. According to Bykov, Orbeli's error was to lead physiology away from the agenda established by Pavlov. But, this constituted an administrative mistake more than a scientific one.

Bykov's first reference to Orbeli was positive and emphasized how his scientific work had helped to combat the notion that the somatic and mental spheres of the brain function independently of one another. The report clearly contained a good deal of criticism of Orbeli, Anokhin, and others, but Bykov did not completely dismiss them as pseudoscientists the way Lysenko had dismissed Dubinin, Zhebrak, and others. Rather than simply ridiculing everything associated with Orbeli, Bykov sought to show that Orbeli's mistakes arose because he "deviated from Pavlov's teachings." Orbeli, he seemed to imply, was terribly misguided in his choice of research subjects but was still a good scientist. His work was of "great interest" and his "theories concerning the sympathetic nervous system and the study of problems pertaining to what is called evolutionary physiology are important in themselves. . . ." Bykov immediately added, however, "but it must be said that they have only indirect bearing on the problems raised by Pavlov himself." Later he commented that "we have no intention of belittling the research of Orbeli's school. . . ." It is hard to imagine Lysenko declaring, as Bykov did at the end of his report, that "I have deep esteem for all our physiologists and biologists. . . ." But to show that Bykov was no Lysenko does not mean that his report was completely free of the acerbic criticism that characterized the postwar scientific discussions more generally.[60]

Some parts of the report suggested opportunism and willingness, based in fear or faith, to twist facts. In telling the history of Pavlov's founding of "materialist physiology," Bykov downplayed the influence of foreign scientists on Pavlov's work by placing it almost exclusively within the context of Russian research. He also read Orbeli and Anokhin's work selectively. According to Bykov, they fell under the influence of Western and bourgeois theories and made no effort to prevent idealism from seeping into Soviet physiology; Orbeli had abandoned the objective methods of Pavlov in favor of foreign, subjective psychology. Finally, Bykov obliquely suggested that Pavlov was in fact a Michurinist and believed in the inheritance of conditioned reflexes.[61] It is not clear whether Bykov believed what he was saying. He may have been simply following Zhdanov's direction or Stalin's editorial comments.

Bykov outlined so many areas in which Pavlov's work was of paramount importance that it was hardly logical to blame Orbeli for not pursuing them all, even with the vast array of institutes under his control. "Pavlov's science of conditioned reflexes," Bykov proclaimed, "has had a powerful influence on fundamental and cardinal problems of biology, medicine, psychology, and philosophy." Orbeli wrongly assumed that since others were working on conditioned reflexes he was free to explore other aspects of the nervous system. Orbeli's personal interests got in the way of his administrative role of developing Pavlov's ideas. He spent too many resources pursuing questions of the "internal organism," rather than following Pavlov's lead in exploring the inter-action between internal and environmental (or external) factors in the func-tioning of the organism. In short, Bykov charged that Orbeli had marshaled the country's resources to answer the wrong questions.[62]

As Bykov would have it, Pavlov's legacy began with a narrow focus on conditioned reflexes and branched out from there into explanations for all animal and human behavior. Specific research agendas had to stay focused on the study of the cerebral cortex, yet legitimate applications of Pavlovian sci-ence were almost limitless. He called for more work to be done to examine Pavlov's contribution to digestion, pharmacology, biochemistry, prophylactic medicine, spa therapy, physical fitness, ecological physiology, and clinical medicine.[63] The time had come for Soviet physiology to catch up with the perpetually growing myth of Pavlov's all-encompassing genius.

Ivanov-Smolenskii's plenary report, delivered the next morning, singled out Anokhin, who as "a disciple of Pavlov's" supported Beritashvili's work and P. S. Kupalov (who had heretofore escaped criticism), for arguing that some forms of animal behavior could not be explained by reference to either conditioned or unconditioned reflexes. Orbeli, in Ivanov-Smolenskii's estima-tion, propagated "psychophysiological parallelism" and overstated the impor-tance of subjective processes in the analysis and treatment of humans. Rather than accepting subjective sources, which were by definition unscientific and idealist, Ivanov-Smolenskii called for Soviet psychiatrists and psychologists to base their work on the objective physiological findings of Pavlov and his disciples. Even speech, human beings' most complex activity, could be ex-plained using the physiological principles outlined by Pavlov in his discussion of the second signal system.[64]

Subsequent speakers usually followed Bykov's and Ivanov-Smolenskii's lead, while others pressed beyond their pseudocollegial tone. Dmitrii Andree-vich Biriukov, a physiologist from Rostov-on-the-Don who specialized in human reflexes, lambasted Shtern in a style reminiscent of the Agricultural Academy session. He accused her of holding idealist positions, in part because she studied the role of chemical factors in the functioning of the nervous system. She had allowed "fanatical cosmopolitanism, politically unprincipled behavior and kowtowing before foreign false authorities" to prevent her from seeing the problems with her work. Borrowing a phrase from Stalin's articles on linguistics, Biriukov argued that Shtern had formed an "Arakcheev-like regime" in her field.[65] The tone of these attacks against Shtern put the rela-tively mild attacks against Orbeli in perspective.

Not everyone concurred with the keynote speakers' description of the prob-lems in the field. Many recognized the need to be politically engaged and ideologically rigorous but criticized the idea that scientists had to resist new approaches and theories, that political or philosophical errors should be equated with scientific errors, or that anyone should be denied the right to confront his or her scientific detractors. Ezras Astratovich Asratian—Party member, corresponding member of the Academy of Sciences, direct disciple of Pavlov's—accepted the "political and philosophical" importance of the meeting and called for a "united front behind Pavlov's teachings." But then, oddly, he earned a bit of applause from the audience when he called for Bykov to be removed from all but a few of his positions.[66] Some speakers boldly defended Orbeli and questioned the positions outlined in the opening reports. Aleksei Dmitrievich Speranskii, who had worked with Pavlov since the early 1920s and joined the Party during the Second World War, defended innovation among Pavlov's disciples. He added that though the cerebral cortex was the most important part of the nervous system, other parts also should be studied.[67] Orbeli's longtime assistant Aleksandr Grigorievich Ginetsinskii asserted that he could not agree with Bykov's claim that Orbeli had in some way diverged from the problems outlined by Pavlov and his school. Though Ginetsinskii defended past research, he also called for Orbeli to commit more of his time and resources to studying the cerebral cortex.[68]

Some of the accused also took the floor, fearful that their whole approach to science was being denounced. The morning session on June 30 opened with speeches by Kupalov and Orbeli. Kupalov, whom Zhdanov had described before the meeting as a leading Pavlovian, defended himself against Ivanov-Smolenskii's accusations that he sided with Beritashvili and, specifically, Beritashvili's efforts to complicate Pavlov's understanding of reflexes. Kupalov feared the rigidity of the opening reports: "is it possible that we, the Russian, Soviet scientific successors of Pavlov and Sechenov have lost our right to create new scientific terms and the right to understand and systematize new facts that we have uncovered? I think that we have not lost that right."[69] Speranskii's, Ginetsinskii's, and Kupalov's counterattacks against Bykov and Ivanov-Smolenskii were brave, given the circumstances.

Orbeli delivered an even more strident defense of himself, his colleagues, and his understanding of proper scientific discourse. He refused to accept the formulation that characterized him as a good scientist who had allowed personal interests to drive the field toward the wrong questions. And he paid a high price. He began perhaps disingenuously by thanking the organizers of the session and welcoming the call for criticism and self-criticism as a way of further developing the Pavlovian tradition. But he was clearly taken aback by what was evidently a carefully organized attack against him. He, like Vavilov, had assumed that the meeting would be scholarly. True to Stalin's tactical advice, the full assault so carefully planned by Zhdanov over the course of the spring of 1950 came as a surprise to Orbeli:

> I, unfortunately, must rebuke the organizers of the session themselves. The point is that if particular individuals are singled out and exposed to rather sharp criticism, then in the event of a free scientific discussion it is extremely important to familiarize

these individuals with the substance of the accusations and criticisms. Even when it is a question of criminals, they are given the chance to read the indictment so that they can defend themselves or say something in their own defense. In this case that was not done and we, the defendants, find ourselves in a difficult situation.[70]

The attacks clearly did not correspond to either his sense of how science was supposed to work or his sense of justice.

Orbeli could admit to some mistakes as a scientific administrator if not as a scientist. He noted that he had taken on too much work and that when he assumed his administrative positions after Pavlov's death, some of his colleagues had "met him with extreme hostility." Orbeli blamed this in part on the fact that while he believed that Pavlov's work on the higher nervous system was his most important contribution to physiology, it was not his sole contribution. Using similar logic, Orbeli defended his choice to study the evolution of the nervous system, arguing that this was a legitimate development of Pavlov's ideas.[71]

Orbeli also asserted that Pavlov never doubted the existence of the subjective world. He accused Bykov of quoting him (Orbeli) out of context in order to make him look as though, in supporting research on both objective and subjective categories, he was somehow falling into a position of psychophysiological parallelism. Continuing with his irreverent style, Orbeli dismissed other accusations that had been raised on the previous day. For instance, he noted that the reports praised Asratian's work without noting that it had been conducted under Orbeli's leadership. He observed that, "on the one hand, all that is useful is attributed to different individuals, as if I didn't exist at all and I was irrelevant, and all that is negative is dumped on me."[72]

Orbeli even defended his positive references to foreign physiologists by stating that he could separate their philosophical views from their legitimate physiological contributions and reminded the audience that his work in foreign countries had been conducted with Pavlov's approval. This logic is reminiscent of the arguments of some physicists who claimed to be able to use Einstein's, Heisenberg's, and Bohr's science without also falling victim to their idealism. Orbeli defended his approach aggressively, stating, "Neither Lenin nor Joseph Vissarionovich Stalin ever said anywhere that because a person takes an incorrect philosophical position you must completely dismiss him."[73]

Some of Orbeli's "self-criticism" seemed disingenuous. He suggested that one of his biggest mistakes was that he had been "ashamed to bother the [Party] leadership . . . with appeals." Timidity, he argued, had prevented him from seeking out the Party's—and Stalin's—advice in how to further Pavlov's scientific legacy. He also explained that the material and intellectual support he had already received as Pavlov's disciple made him apprehensive about approaching Stalin to ask for anything more. Perhaps he calculated that by admitting political and administrative errors he could shield himself from attacks against his science. In any case, Orbeli's contentious speech was met with applause from the audience.[74]

If Orbeli expected the support of his colleagues, however, he must have been gravely disappointed in many of the subsequent speakers. Even those who had worked closely with him claimed in their speeches to be taken aback by the tone of his speech and his lack of genuine self-criticism.[75]

The philosopher Georgii Aleksandrov, who might have been expected to have some sympathy for an academic administrator raked over the coals by one-time colleagues and subordinates, enthusiastically joined the bashing. While Aleksandrov also criticized Anokhin and Beritashvili, his main attacks centered on Orbeli and his philosophical views. He accused Orbeli of down-playing Pavlov's commitment to materialism and his deep concern with philo-sophical questions. In Aleksandrov's creative analysis, Pavlov was committed to attacking idealism and subjectivism and was a direct descendant of Russian classical materialists of the nineteenth century. Aleksandrov reminded the audience that scientists were involved in the great philosophical battle to defend materialism from bourgeois and idealist attacks. No deviation from this ideological commitment could be tolerated, especially among leading physiol-ogists. Aleksandrov compared the session with previous discussions in philoso-phy, biology, and linguistics and noted that Stalin and the Central Committee had organized each of them in order to ensure that science would attain the level necessary for building communism.[76]

Criticizing Orbeli was not the session's only goal. Two of the other targets of attack, however, were not in attendance. Beritashvili never showed up, despite the efforts of the organizing committee.[77] Shtern was in prison awaiting trial along with other members of the Jewish Anti-Fascist Committee. Most members of the committee were executed; Shtern received a sentence that combined prison and internal exile to Dzhambul, Kazakhstan.[78] Some of Beri-tashvili's and Shtern's colleagues, however, spoke in their absence. Shaken by the anticosmopolitan campaign, D. I. Shatenshtein chose not to defend Shtern, his former scientific collaborator. He accepted that the criticism of her was "completely justified." "Shtern's scientific positions," he continued, "should be qualified as methodologically dishonest and scientifically defec-tive." Shatenshtein also admitted his own errors in supporting Shtern in the past, a situation that he attributed to a lack of scientific criticism. Even such model self-criticism was not enough to satisfy the needs of the session, how-ever. Bykov later pronounced Shatensthein's speech "strange" because, though he "came forward as a repentant sinner," he did not give a detailed scientific explanation of Shtern's mistakes or ridicule those of his coworkers who had yet to come forward to denounce Shtern's work.[79]

Beritashvili fared a little better in absentia. Undeterred by criticism, one of Beritashvili's colleagues from the Institute of Physiology of the Georgian Academy of Sciences, N. N. Dzidzishvili, tried to defend their contribution to the development of Pavlov's work. While accepting that academic discussions were necessary for science to advance, Dzidzishvili called for productive debate. "Criticism should be directed toward the clarification of scientific truth, and not toward groundless accusations and disparaging new discoveries based on inaccurate quotations taken out of context and placed in front of many exclamation points and question marks." Needless to say, Dzidzishvili's speech did not conform to the goals of the session. When given the chance to rebut him, Bykov insisted that the speech was "incomprehensible" and "childishly naïve" about the seriousness of the situation and the grave danger of "conservative views seeping into our country from abroad."[80] Still, Beritashvili and Shtern commanded relatively little attention when compared to Orbeli and Anokhin.

For some speakers, Anokhin's sins were of greater consequence than Beritashvili's, because Anokhin hid behind "a mask of belief in his teacher [while] systematically and persistently attempting to revise his teachings using the rotten position of pseudoscientific and idealist 'theories' of reactionary bourgeois scientists."[81] Having witnessed the negative reactions to Orbeli's speech, Anokhin made what seemed to be a serious effort at self-criticism. He admitted to a full range of mistakes, confessing that his account of the history of conditioned reflexes was "absolutely incorrect" and that he should have paid more attention to the ideological and political implications of his statements. He now saw that he had overvalued the accomplishments of foreign physiologists and neurologists and had been insufficiently critical of Beritashvili. Anokhin freely admitted all these errors, but he also put them in the category of "ideo-political" mistakes. He hoped that in the future he would give concrete assistance to the "common struggle of Soviet scientists against reactionary, foreign theories." But he complained that there were other genuine problems in physiology that were not being addressed at the session and that in general there was little exchange of opinion among physiologists. Still, the overall tone of his speech was humble and self-critical.[82]

Orbeli must have noticed the humility of Anokhin's speech and the lack of support his own combative effort had received from his colleagues. Recognizing that the meeting was indeed stacked against him, on July 4, the last day of the session, Orbeli once again took the floor. Still he saw his errors as primarily political and not scientific. He claimed that he "immediately understood the erroneous and dissatisfactory nature" of his first speech and was so impressed by the criticism he heard that he felt the need to speak again. He admitted that the first speech "lacked tact and was politically inappropriate," especially in his references to criminals and defendants. He blamed his error on his inexperience with receiving criticism and his lack of preparation. Specifically, he accepted that as a leader of Soviet physiology he had not created an environment for criticism and self-criticism and had not worked hard enough to develop certain areas of Pavlov's science, including the second signal system. He also now claimed to see that some of his articles and speeches did not accurately portray his understanding of the importance of the higher nervous system. He took Aleksandrov's criticism of his philosophically prob-lematic formulations to heart, but in defense of his 1908 doctoral dissertation, which contained "idealist elements," he reminded the audience that Pavlov had approved it. He also recognized that he had not done enough to struggle against the Soviet and foreign critics and revisers of Pavlov's work. In sum-ming up, Orbeli told the audience that he hoped to learn from the "business-like and comradely" criticism of the session.[83]

In their concluding speeches, Ivanov-Smolenskii and Bykov continued their polemics almost as if Orbeli had not already recanted. Likewise, they took Anokhin and others to task for insufficiently recognizing their errors and for trying to defend their mistaken views. Bykov reiterated that physiologists still did not fully understand the broader context in which the session was taking place: "the successful development and propagandizing of Pavlov's teachings are an extremely important part of the ideological struggles. Only this can explain why the reactionary statements of foreign scientists and pseu-doscientists have not been met here with militant, sharp, Party criticism."[84] Noting that there remained a whiff of cosmopolitanism in the air, he quoted a Mikhaelkov poem:

> There are still some who at their table
> all things Soviet with scorn abuse
> And with oily admiration ooze
> For all that bears a foreign label,
> Yet Russian bacon can't refuse![85]

Bykov declared the session a success, claiming that the value of scientific discussion had been proven once again. He also noted, with pride, that the whole country had been following the proceedings and that the people "love science, are interested in it, and are concerned about its fate just as we are." Vavilov closed the session with generalizations, stating that the session would have a "decisive influence on the further history of Soviet physiology and biology in general." Responding to one participant who was concerned that so many of the discipline's prominent scientists had been criticized and that the session had revealed the horrible state of Soviet physiology, Vavilov em-phasized that much had been accomplished. In light of the excellent condi-tions for science created by the Soviet government, he declared that the future of the field looked bright. With obligatory praise for Stalin, the session was officially closed.[86]

The aftermath of the Pavlov Session shows the administrative power of the Science Section in full force. First, the Party passed the usual high-level resolutions. Ten days after the close of the Pavlov Session the Politburo accused Orbeli of holding an "intolerable monopoly" in physiology that "contradicted the spirit of Soviet science and interfered with its free development." A resolution also singled out Anokhin for "serious errors and distortions of Pavlovian science." In order to eliminate competing scientific schools, the Politburo required the Academy of Sciences and the Academy of Medical Sciences to take measures to unite physiologists behind the further development of Pavlov's teachings.[87]

Specifically, Orbeli lost his posts as the director of the Institute of Physiology of the Academy of Sciences and the Institute of Evolutionary Physiology and Pathology of the Higher Nervous System of the Academy of Medical Sciences. These two institutes joined with Bykov's institute to form the single Pavlov Institute of Physiology under the Academy of Sciences. Asratian and Ivanov-Smolenskii became director and vice director, respectively, of yet another new academy institute, this one for the higher nervous system. Anokhin lost the directorship of his institute in the Academy of Medical Sciences. Later resolutions also fired Orbeli as editor of the *Physiology Journal of the USSR*, as a member of the editorial boards of *Achievements in Modern Biology*, the *Herald of the Academy of Sciences*, and the Soviet science magazine *Nature*. In his place, the Central Committee appointed scientists closely aligned with Bykov and Ivanov-Smolenskii and created a new journal, the *Pavlov Journal of Higher Nervous Activity* with Ivanov-Smolenskii appointed editor in chief.[88] All these moves resemble those actions taken by the Politburo after previous discussions. The last point in the Politburo resolution, however, created something unprecedented: a "Scientific Council on the Problems of the Physiological Teachings of I. P. Pavlov."[89]

The Scientific Council had three main responsibilities: overseeing work in physiology by receiving reports from the major physiology institutes in the country; organizing yearly "businesslike and critical" discussions of Pavlovian physiology; and presenting annual reports directly to the government and the presidium of the Academy of Sciences on the work of physiology institutes. Bykov chaired the Scientific Council; other members included the more strident Ivanov-Smolenskii and the former Orbeli colleagues Asratian and Kupalov.[90] The council met eight times over the next two years to review work in the field. This was the only time in the postwar period that the Central Committee created a standing administrative committee to ensure that the results of a particular discussion were properly enforced.

On July 22 the Central Committee ordered Zhdanov and Kruzhkov, the deputy head of Agitprop, to take the "necessary measures" to ensure that all relevant ministries and academic institutions worked to put into effect the conclusions of the session. By August they had planned thirty-nine specific changes to be made, including further reorganization of various institutions, redesigning research and teaching plans, preparing publications, and encouraging new graduate students in Pavlovian science. They spelled out specific instructions for the Academy of Sciences, the Academy of Medical Sciences, the Ministry of Health, the Ministry of Education, the Ministry of Agriculture, the Ministry of Film (to prepare films about Pavlov's accomplishments), the Ministry of Foreign Trade (to secure monkeys for research), and the Ministry of Enlightenment of the RSFSR.[91]

Stalin, Suslov, and Malenkov forwarded questions concerning the Pavlov Session and its results to Zhdanov, who took responsibility for seeing that the decisions were implemented.[92] Pavlovian physiology was the only legitimate physiology in the Soviet Union, and Agitprop was determined to enforce that line. According to Kruzhkov and Zhdanov, Bykov was the primary interpreter of Pavlov's work. When the minister of health clashed with Bykov, Agitprop scolded the minister, not the physiologist.[93] When a scientist pointed out Bykov's past mistakes, Zhdanov dismissed the charges, noting that Bykov had already undergone the appropriate self-criticism.[94]

Others who had risen to the top of the scientific establishment in the wake of the discussion were less secure. Airapet'iants, who had been the secretary of the session's organizing committee and in July 1950 was appointed Bykov's vice director of the Institute of Physiology, was unceremoniously removed from his post in the middle of 1951. Basing their decision in part on a Ministry of State Security (MGB) report that noted that he was "an unprincipled person" and a "careerist," Agitprop concluded that Airapet'iants had hired people who were neither scientifically nor politically reputable. Interestingly, the MGB report accused Bykov of the same thing, but even though he was director of the institute in question, the Science Section failed to reprimand him.[95]

Bykov survived another MGB memo that circulated in the Central Committee. On February 18, 1951, the Leningrad regional office of the MGB sent the regional Party boss a top secret memo about a conversation Bykov had evidently had with a colleague the month before. The memo made its way to Malenkov, who within two days forwarded copies to the secretaries of the Central Committee as well as Zhdanov. The memo quoted Bykov as saying:

> They've nominated me to be a deputy of the Supreme Soviet of the RSFSR and now I don't see an end to my speeches, meetings with Party organizations and executive committees and also to filling out questionnaires. I don't understand why they have to ask me questions and make me fill out questionnaires when everything about me is already known.
>
> In our country there is one person who cannot be criticized (the leader), but others are either raised up higher than they deserve or they are crushed. Is it really possible to work charitably and productively in these conditions. I have said openly many times, that in these conditions I. P. Pavlov, Lebedev and other world-famous scientists could not have displayed all of their creative abilities. Not long ago I was in Bucharest. I lived in King Mikh's [sic] castle. Talking and meeting with Rumanians convinced me that their material and spiritual culture is much greater than ours.

The influence of French and German culture is still felt there. They still have not wasted this influence and it benefits them. Among them I felt like a human being. That pleasant environment was so beneficial that I felt healthy and did not even worry about my bowels. That's sanity! I returned here and the depression, exhaustion, irritation, and troubling atmosphere began again.[96]

The report was damning in a number of ways. It showed Bykov harboring "cosmopolitan" tendencies and revealed how uncomfortable he was with his appointed post as the country's heir to Pavlov. It also suggested that the USSR was not creating an environment in which scientists could be productive. The Central Committee did not take any action on the MGB memo, but it kept the compromising information on file, as was common practice. Whether or not this report was accurate, it certainly would have proved effective if Bykov in any way openly lapsed in his pursuit and punishment of Orbeli and other so-called anti-Pavlovians. It appears that Stalin's instincts about Bykov had been right: he was "a little timid" and did not like "to mix things up." Zhdanov kept a careful eye on him. In a memo to Malenkov in April 1952, Zhdanov noted with praise that Ivanov-Smolenskii was combating "unprincipled efforts by Academician Bykov to 'make peace' with people holding mistaken views in physiology."[97] But as long as Bykov executed the decisions of the Pavlov Session, Zhdanov accepted him as the main representative of Pavlovian physiology.

As a fellow enforcer of the Party line, Ivanov-Smolenskii seems to have shared Bykov's protection from criticism. In March 1952, the Central Committee received a series of letters about the Institute of Higher Nervous Activity, where Asratian was the director and Ivanov-Smolenskii the vice director. In a letter to the Party secretary Nikita Khrushchev, a scientist from the institute complained that Ivanov-Smolenskii was running an "Arakcheev regime," intentionally misinforming the Science Section about work at the institute, and erroneously accusing Asratian of anti-Pavlovian ideas. Other scientists from the institute and even the secretary of the institute's Party organization sent similar letters to Malenkov around the same time. Malenkov and Khrushchev forwarded the letters to Zhdanov, asking him to figure out what was going on and to report back to the Secretariat.[98] Zhdanov's report to Malenkov unambiguously took Ivanov-Smolenskii's side in the dispute. In contrast with the letters' assertions, Zhdanov argued that Asratian had "not been able to free himself from the weight of Orbeli's mistakes" and had used his position as head of the institute to hire many of Orbeli's students. The effort by Asratian's colleagues to get Ivanov-Smolenskii fired was "contrary to the interests of the institute and contrary to the interests of the further development of physiology in our country." Zhdanov recommended that the presidium of the Academy of Sciences fire Asratian instead, Malenkov concurred, and by June they had removed him.[99]

Figure 12. Even after his father's death in 1948, Yuri Zhdanov thrived at the head of the Central Committee's Science Section. He married Stalin's daughter, Svetlana, in 1949. Here Yuri and Svetlana are shown voting together in February 1951. They separated in early 1952, but Stalin remained pleased with Zhdanov's work on ideology and science. Courtesy of RGAKFD.

While Zhdanov smoothly brushed aside Bykov's and Ivanov-Smolenskii's potential problems, the pressure on Orbeli, Beritashvili, and Anokhin only increased. The "Scientific Council on the Problems of the Physiological Teachings of I. P. Pavlov" proved to be an efficient instrument of control. One physiologist likened it to an inquisition, "since all of its work consisted of challenges and interrogations of various physiologists."[100] Its purpose was primarily punitive, not productive.

Even after losing many of his most important positions in July 1950, Orbeli was not completely destroyed as a scientist. He remained a member of the academy presidium, a member of the editorial board of a relatively minor journal, and, most important, the head of his original department at the Lesgaft Institute in Leningrad. With a small group of assistants, he set about trying to develop research strategies that would satisfy himself and the Scientific Council. As his biographer noted, Orbeli maintained his trust in the Party and in Stalin and believed that with diligent attention to the classics of Marxism-Leninism and of Soviet physiology (including, of course, the stenographic record of the Pavlov Session) he would be able to understand and overcome his mistakes.[101]

For its part, the Scientific Council had little interest in rehabilitating those who had been unmasked as anti-Pavlovian at the session. In December 1950, Speranskii and Kupalov presented their research to the council, which criticized them for not fully comprehending past errors. Both speakers "accepted the correctness of the criticism" and agreed to resubmit their proposals. The next meeting offered more of the same. Beritashvili, who remained the director of the Georgian Academy's Institute of Physiology and the head of the Physiology Department of the Stalin University in Tbilisi, presented his ideas to the council. Bykov and the others accused him of being an "open dualist" with "pseudoscientific," "reactionary" theoretical concepts and of leading an "Arakcheev regime" in Geogian physiology. The president of the Academy of Sciences reported to Malenkov that the meeting had been run "in the spirit of free discussion and sharp criticism." Zhdanov agreed and supported the proposal that a group of physiologists be sent to Georgia to organize a "wide discussion of Beritashvili's mistakes."[102] Six months later, the Scientific Council reported back to Malenkov that their committee had gone to Tbilisi and found that scientific organizations there had "actively propagandized Pavlov's ideas and criticized the idealist conceptions of Beritashvili." Still, Beritashvili remained tied to his "dualist conceptions."[103]

Anokhin met with a similar fate when he went before the Scientific Council in November 1951. Despite his efforts at self-criticism the council decided that his pronouncements were superficial. The members of the council pointed out Anokhin's "core methodological errors" and parts of his scientific plan that were anti-Pavlovian. As Zhdanov interpreted the meeting, the council "showed him [Anokhin] the path to reforming his scientific work."[104]

The Scientific Council's untiring commitment to crushing its opponents was most evident it its dealings with Orbeli. At the end of 1950, Orbeli may still have believed he could rebuild his academic reputation. He wrote a letter to Molotov that earned him an invitation to discuss his situation at the Central Committee. But as the Agitprop bureaucrat who met with him reported to Molotov, Orbeli still insisted that he did not make idealistic or anti-Pavlovian mistakes and that criticism of him was unjust. The memo did not recommend giving Orbeli additional assistance for his lab.[105] In early June 1951, the Scientific Council met to discuss Orbeli's research plan for his laboratory. Orbeli's report was full of praise for dialectical materialism as the one true philosophy, for Stalin and his leadership in science, and for the benefits of critical discussion. He also admitted to the "seriousness and correctness of the criticism" he had received. He accepted that his de-emphasis of the higher nervous system, his willingness to support genetics research, the introduction to his 1908 dissertation, and his lack of clarity in discussing subjective and objective categories in his work were all grave mistakes. If he thought that admitting his errors would gain him the right to discuss once again the merits of his scientific plans, he sorely underestimated the vindictiveness of his opponents.[106]

Instead, for two days Bykov, Ivanov-Smolenskii, and others bombarded him with antagonistic questions about his philosophical positions, his attitude toward his past errors, and the degree to which he had really reformed. The discussion was heated, and Orbeli seemed to lose his temper occasionally as his opponents kept returning to the same issues, such as his dissertation or his support of genetics, over and over again. Two examples of the type of criticism he faced should suffice to give a sense of the meeting. After the Pavlov Session, Orbeli decided that to continue studying the sympathetic nervous system made little sense, especially given the session's emphasis on the higher nervous system. So, in his research plan, he set out to study the higher nervous system as it developed in children, since his laboratory was part of the Academy of Pedagogical Sciences. But Bykov saw even this attempt to return to the foundations of Pavlov's science as untrustworthy. He wanted to know why Orbeli had abandoned the study of the sympathetic nervous system rather than studying it from a Pavlovian point of view. Clearly, Bykov concluded, Orbeli was trying to avoid having to reformulate his non-Pavlovian understanding of physiology and was running away from his mistakes, rather than correcting them. Thus, his whole scientific plan was deemed unacceptable.[107]

A similar logic prevailed when Orbeli tried to explain his understanding of the role of the subjective and objective in physiological research. Even as he decried dualism and declared himself at one with Pavlov on the supremacy of objective over subjective research, the council members criticized him for even comparing the two concepts. To those determined to see errors in everything he said, Orbeli was clearly implying that Pavlov's objective method required confirmation from the subjective methods of psychologists. This meant Orbeli did not believe Pavlov's method had been proven yet; Orbeli was thus an idealist, under the sway of bourgeois pseudoscientists.[108]

The official report from the Academy of Sciences to Malenkov noted that participants in the meeting were unanimous in their criticism of Orbeli and "in this respect they took a strong position in the struggle for Pavlov's materialist teachings." Orbeli "admitted the correctness of the criticism and declared to the Scientific Council his desire to stand on the road to Pavlovian science." Cynically, they noted that he had made similar promises at the Pavlov Session. In reporting to Malenkov on the meeting, Zhdanov assured him that it had been handled properly and that Orbeli had tried to defend his mistakes, discredit Pavlov, and dishonestly profess his agreement with the criticism he received.[109]

In December 1952 the Scientific Council met once again to discuss Orbeli's work. In anticipation of the meeting, Orbeli wrote a long letter to Stalin accepting the results of the Pavlov Session but noting that the persistent criticism he faced from the Scientific Council was not justified. His cover letter stated that he found himself in an "extremely difficult situation" that he could not escape from without Stalin's help. Stalin's secretary simply forwarded the letter to Zhdanov, the man behind the anti-Orbeli campaign.[110] Zhdanov reported back to Malenkov, who then circulated this report to all the members of the Secretariat, that Orbeli maintained "subjectivist-idealist views in science," "continued to discredit I. P. Pavlov," and still was not correctly developing Pavlovian physiology.[111] Meanwhile, some of Orbeli's colleagues sent a letter on his behalf to Malenkov, who passed it on to Zhdanov. The head of the Science Section reported back that the letter simply provided additional evidence that Orbeli and his group had not "exposed Academician Orbeli's errors in physiology [and that] in the collective there is not the necessary amount of criticism and self-criticism of their scientific work."[112]

Needless to say, the December 1952 meeting of the Scientific Council did not go well for Orbeli. He was probably correct to assume that only Stalin's intervention could have saved him from further humiliation and setbacks. It was not forthcoming. Orbeli once again met an antagonistic audience led by Bykov. The resolution from the meeting reasserted that Orbeli continued to hold anti-Pavlovian positions and had not used past criticism from the Pavlov Session or his previous meeting with the council to reform his methodological mistakes. Furthermore, the Scientific Council decided to inform the Pedagogical Academy of the "unsatisfactory situation" in Orbeli's only remaining laboratory. Evidently, Bykov believed that his fellow physiologists had not been vigilant enough in their criticism of Orbeli. The resolution noted that even the members of the council itself had not worked hard enough to expose Orbeli's idealist positions. With that in mind, the council's resolution recommended that editors of the country's "physiological, biological, medical and pedagogical journals systematically publish articles exposing the harm inflicted by Orbeli's . . . idealistic, subjective method."[113]

From the perspective of Agitprop it is hard to imagine a more successful discussion than the Pavlov Session. Not only was Pavlov easily turned into an ideal comrade-scientist; Zhdanov also found in Bykov and Ivanov-Smolenskii scientists who obediently followed their assigned roles in the national discussion and in the administrative organ created to enforce its conclusions. For the Central Committee bureaucracy the discussion in physiology accomplished three things that had proved elusive in the other discussions. First, Agitprop and the Science Section were able to run the discussion smoothly and without massive repercussions for themselves. After the philosophy discussion, by contrast, the head of Agitprop was dismissed. A year later Yuri Zhdanov seriously

miscalculated his role in the biology discussion and found himself at odds with
Stalin. The physics discussion failed. The linguistics discussion was resolved
only by Stalin's own intervention. But in physiology Zhdanov steered events
from start to finish without any serious missteps, in no small part because he
consulted with Stalin as his plans progressed. The intensity of the work clearly
affected Zhdanov's personal life, however. At the beginning of 1952 his wife,
Svetlana, wrote to her father, Stalin, complaining about how Zhdanov acted
like a "dry professor," a "heartless scholar" who was so concerned with his
books that he had no time for his family. Stalin allowed them to divorce but
continued to support Zhdanov's work in the Central Committee.[114]

Zhdanov avoided pitfalls in his work by creating a stable compound of science and ideology. This constituted Agitprop's second accomplishment. By
creating and enforcing the myth of Pavlov as a patriotic, materialist, and all-
encompassing scientist, Zhdanov and others were able to merge two powerful
strands of postwar Soviet ideology: patriotism and doctrinal orthodoxy. Zhdanov used Pavlov to show how Soviet physiology was distinct from Western
physiology and, perhaps more important, how it was philosophically consistent with dialectical materialism. Pavlov's science was not explicitly labeled
Party-minded science. Rather, Marxist-Leninists shared the fundamental assumptions of Pavlov's scientific methodology—that the world can be reduced
to concrete phenomena that scientists can measure and utilize for the benefit
of man.

Finally, in physiology Agitprop ran a discussion where the heavy hand of
the Party, though always felt, was never visible. The discussion was scientific
in that it took place without the open participation of anyone in the Party
leadership. People referred to Stalin's support and guidance and quoted him
often. But no one represented him, or anyone in the Party's leadership, as
an expert in physiology. The creation of the Scientific Council encouraged
scientists to monitor themselves. No doubt, having incriminating evidence
on file about Bykov and Ivanov-Smolenskii tightened Zhdanov's grip on the
levers. But, in the end, scientists reprimanded one another at the session and
controlled one another afterward. Whether they did so out of fear or filial
humility, ambition, or spite cannot be known from the records. Subsequent
events show, however, that whatever the motivation, for many it did not survive Stalin and the subsequent removal of Zhdanov from the Science Section.

In May 1953 Lina Shtern returned from exile in Kazakhstan, part of a
broader amnesty after Stalin's death. The next year the academy discredited
and disbanded the Scientific Council. In 1955, scientific colleagues greeted
Orbeli's appearance at a meeting of the Physiology Society with an ovation
that lasted minutes. In 1956, the Academy of Sciences upgraded his laboratory
to an institute. Before his death in 1958, Orbeli received the Order of Lenin
for outstanding service to the Soviet state.[115]

Though scientists and the Party did not renounce the Pavlov Session itself, the post-Stalin retreat from the calcified definitions of physiology suggests that under the surface, even during Stalin's time, there was a growing sense of cynicism about the Party's role in science. Bykov's laments about the system and apparent desire to "make peace" with his opponents (as Zhdanov put it) suggest that even those scientists in charge of administering the Party line had their doubts. Whatever the case, soon after Stalin's death, the dismantling of the rigid interpretation of Pavlov's legacy began in earnest.

CHAPTER 7

"EVERYONE IS WAITING"

Stalin and the Economic Problems of Communism

Political economy—the subject of the last Central Committee–sponsored scholarly discussion of the Stalin period—served two related and essential functions for the Party during the early Cold War. First, it provided the scholarly foundation for the Marxist-Leninist critique of capitalism. Second, it buttressed Soviet proclamations to the world about the advantages of socialism. As Stalin said, "Political economy is serious work."[1] Ideally, political economists were both scholars and political agitators. Like philosophers, political economists balanced their "academic" and "objective" scholarship with the Party's demands for politically relevant and popular work. In 1946 and 1947 Stalin and Andrei Zhdanov had chastised philosophers for being too distant from politics and for shying away from their mission as "publicists" in the battle with Western philosophers along the "ideological front." In five meetings with economists that began in 1941 and continued through 1952, Stalin took a different tack. He tried to remove "propaganda" from economists' work and emphasized the "scientific" basis of their field.

Like their colleagues in other disciplines, economists found themselves caught between their discipline's approach to academic problems and the demands placed on them by the Party and Stalin. As in other fields, Marxist-Leninist theory left no room for contradictions between scientific conclusions and Party decrees. In these circumstances, discussing the political economies of capitalism and socialism often led to questions that economists were reluctant to address without assistance from political authorities. Even when Stalin emphasized that economic laws, "like the laws of natural science," were beyond the power of governments or individuals to create or destroy, economists looked to Stalin to provide guidelines for identifying those laws.

The parallels between the political economy discussion and the previous scholarly discussions go beyond the tension between Party-dictated truths on the one hand and scientific truths on the other. Structurally, the last discussion shared elements of those that preceded it. As with the philosophy discussion, what began as a meeting concerning a specific book turned into a more general appraisal of the state of a whole academic discipline. As was the case with physics, the political economy discussion was never publicized, and the minutes of important meetings remained hidden in the archives, obscuring the meaning of those aspects that did make their way into the press. Similarly to what happened in the linguistic discussion, Stalin suddenly and surprisingly intervened, altering the work of professionals in the field. Like the biology meeting, the political economy discussion was intertwined with internal struggles at the top of the Party hierarchy. In addition to Stalin, Politburo members Andrei Zhdanov, Georgii Malenkov, and Nikolai Voznesenskii were all directly involved in decisions about political economy, and their political fates were closely connected with the matter.

The centrality of political economy to the Soviet Union's raison d'être meant that the stakes in this field were higher than in any of the others. The whole purpose of the Soviet state and Communist Party was to build the economic foundations for communism. Yet there were no acceptable Soviet textbooks on the socialist economy or the transition to communism. Stalin commanded economists to fill this void and commissioned a political economy book that would provide the canonical text for the socialist system. This involved updating the classics of Marxism-Leninism for the contemporary world by reconciling doctrine with Soviet reality. The Second World War, the onset of the Cold War, and the development of "people's democracies" in Europe all suggested a need for reevaluating Soviet ideas about capitalism and socialism. More than other academic disciplines, political economy was fundamental to the workings of the Soviet system and the process of presenting that system to the world. Recognizing the significance of the political economy textbook, Stalin met with prominent economists, participated directly in their discussions, and addressed economic problems of socialism in his last theoretical work.

Even within the confining intellectual environment of postwar Stalinism, some significant debates among economists did emerge. In meetings at the Institute of Economics, in economic publications, and in discussions about the content of the much anticipated official textbook on political economy, economists voiced disagreements about the role of the state in modern capitalism, the role of economic laws in socialism, the process of distribution of commodities under socialism, and the relation of the USSR to other nations.

Their tasks were to articulate the theoretical foundations of the Soviet system, the role of the USSR as a leader among socialist countries, and the theory behind the growing antagonism between the USSR and the United States and its Western European allies. Over time, however, debates on these topics were silenced as economists measured the costs of committing themselves to ideas that had not been endorsed by the Central Committee. Without the political authority to make definitive declarations on economic questions, the field stagnated. In this context, writing a political economy textbook "approved by the Central Committee" became the overriding priority of Soviet economics and eventually the catalyst for encouraging Stalin to enter economic debates.

Given the importance of postwar economics for understanding the history of the USSR and of the Cold War, it is not surprising that Western scholars gave considerable attention to this aspect of the late Stalinist scholarship.[2] Still, in comparison with the other scholarly debates of the period, the political economy discussion has not been the subject of much recent, archive-based research.[3] What emerges from the newly accessible documentation is a story of an academic discipline struggling to meet both scholarly standards and Party demands. At times certain scholars appeared to have struck the proper balance. But no sooner had they established their authority in both the scientific and the political sphere than they were ruthlessly attacked. There was no equivalent of Pavlov for political economy. Marx, Engels, and Lenin provided the canonical ideas in the field, but no one knew how those ideas could be modernized to make sense in the postwar world. Nor was there an equivalent to Lysenko: no contemporary economist managed to combine scientific arrogance, practical promises, and political opportunism with such stunning success. In this discipline, there was room for one comrade scientist—Joseph Stalin.

Like the other great scholarly debates of the postwar period, the political economy discussion had its origins in the late 1930s, when the Central Committee set about systematically legitimizing the state and Party's actions since the Revolution. This entailed the codification of policies that in practice were chaotic and only weakly grounded in doctrine. *History of the Communist Party of the Soviet Union (Bolsheviks): Short Course* was clearly the primary text in this campaign. Published in 1938 with a note proclaiming "Edited by a Commission of the Central Committee" and "Authorized by the Central Committee," the *Short Course* cast Stalin as the only legitimate heir to Marx and Lenin. It presented Stalin as Lenin's closest comrade in the Revolution and Civil War, and as the driving force behind subsequent events in Soviet history, from collectivization and industrialization to the adoption of the 1936 Constitution. The ultimate source for the praise was clear: Stalin wrote the chapter of the *Short Course* titled "Dialectical and Historical Materialism" and carefully edited the rest of the book.[4]

"Short courses" on other subjects were planned as well, with the Central Committee commissioning two textbooks on political economy in 1937. Stalin was harder to please on this topic, however, and for the next sixteen years the leader repeatedly met with economists in an effort to forge a textbook on political economy that could provide a basis for educating Soviet cadres and communists around the world. There was an obvious difference between the goals of the *Short Course* on the history of the Party and any textbook on political economy. A major point of the *Short Course* was to glorify the Party and to closely intertwine Stalin's personal biography with the great events of Soviet history. Economists, in contrast, were supposed to identify and describe underlying economic principles that showed the inevitable victory of communism. The *Short Course* set about codifying the past; the political economy textbook had the more difficult task of codifying the future.

Throughout the 1930s, economists had essentially subordinated analysis of the economy to blind praise for political decisions. By definition, whatever Stalin and the Party did was economically sound, so that the economists' role had been simply to rationalize the leaders' decisions using the vocabulary and rhetoric of Marxism-Leninism. Economists now set about systematizing that approach in a course for students, but without challenging the primacy of the Party and Stalin as portrayed in the *Short Course*. Needless to say, the textbook on the political economy of socialism proved difficult to write.

The Central Committee's 1937 decree called for two textbooks on political economy, one for an introductory course edited by Lev Leont'ev, a thirty-six-year-old economist who had joined the Party in 1919, and the second for more advanced students edited by Konstantin Ostrovitianov, the forty-five-year-old director of the Academy of Sciences' Institute of Economics who had joined the Party in 1914. Leont'ev began sending Stalin drafts in 1938. Stalin would comment on them and demand revisions. Stalin also solicited other economists' comments, corrections, and opinions on drafts.[5] The signing of the Nazi-Soviet nonaggression pact, the occupation of the Baltic countries, and the Winter War with Finland did not seem to distract Stalin from his concern with the textbook. In early 1940 Stalin was still not completely satisfied, although he indicated that the final version would be published with the message "Approved by a Commission of the Central Committee." Leont'ev finished another draft at the end of 1940, and on January 29, 1941, Stalin summoned him, along with Ostrovitianov and four other economists, to the Kremlin. Also in attendance were some of the country's most powerful leaders, including Zhdanov, Molotov, Voznesenskii (chairman of the State Planning Commission [hereafter Gosplan] and soon to be candidate member of the Politburo), and Aleksandrov (the up-and-coming philosopher and newly appointed head of Agitprop.)[6] Stalin clearly gave broad political significance to the completion of the textbook.

The meeting lasted an hour and a half. During that time, Stalin criticized the tone and content of the book and expounded on the "law of value" in the USSR, the reasons for planning the economy, wages under socialism, the need to expose the false claims of fascist economics, and more. Stalin had read the textbook carefully and used the meeting with the authors to lay down some general principles about political economy that he felt had been ignored. He spent almost the whole meeting addressing the section on socialist economy, which he believed was still in need of substantial revision.

According to Stalin, the textbook misrepresented the purpose of economic planning. Leont'ev portrayed planning as allowing the USSR to "surmount the law of value" and eliminate the anarchy of production characteristic of capitalism. Planning entailed the coordination of the Soviet economy in keeping with the task of building socialism. Stalin's response: "This is all nonsense, some sort of schoolyard bumbling! Marx and Engels wrote from afar, they should have spoken about contradictions. But why the devil are you feeding us those kinds of abstractions?" For Stalin, planning had more concrete purposes: "The main task of planning is to ensure the independence of the socialist economy from capitalist encirclement. This is absolutely the most important task." Planning allowed the Soviet economy to bypass questions of profitability in the name of heavy industry, in order to provide for the country's defense. Second, planning helped destroy the forces in Soviet society that might give rise to capitalism. Finally, planning helped counter problems of disequilibrium in the economy. In conclusion, he added, "You need to show something new to readers and not endlessly repeat the correlation of productive forces and production relations. This offers nothing. You don't need to praise our system too much and describe accomplishments that don't exist." Leont'ev had written a textbook describing Soviet planning in abstract terms. Stalin wanted a textbook that addressed the practical problems facing the state.[7]

Stalin's criticism of the textbook's description of wages under socialism reflected a similar concern. He pointed out, "The textbook fails to show that people do not work because Marxists are in power or because the economy is planned, but because they have an interest in working. We cling to our interests." In Stalin's view, "workers are not idealists"; they work because of bonus systems and incentive plans. "Get the people interested and people will push ahead, will raise their qualifications, will work better. They clearly understand what earns them more." Since Marx and Engels wrote with "total communism in mind," their writings were not necessarily helpful for describing wages during the period of transition. He reminded the group of politicians and economists in his office, "We have yet to get socialism in the flesh and blood, and we still need to put socialism right, still need to distribute according to labor as is necessary." Plans to distribute according to need had to be delayed. Putting things rather crudely, he added: "We have dirt in the factories and want to go directly to communism. And who will let you in? They are buried in rubbish but desire communism. Two years ago in one major factory they were raising hens and geese. What good is this? Dirty people are not permitted into communism. We need to stop being pigs! They talk about being let into communism. Engels wanted to go directly into communism and got carried away."[8] Wages, then, were not to be leveled, even in theory, and there was no need to apologize for the incentives put in place to encourage worker productivity. It seemed that socialism, at its present stage, remained a long way behind mature communism as it had been outlined by Marx and Lenin.

If the classics did not provide the starting point for understanding the political economy of socialism, what did? Stalin's answer: Soviet reality. In his descriptions of planning and wages, Stalin preferred practical observations of Soviet reality to "abstract" theories. Theorizing resulted from economists' tendency to use Marx and Engels as guides to understanding socialist economics. Stalin claimed to dislike this approach, emphasizing instead the importance of Soviet experience: "If you search for the answer in Marx, you'll get off track. In the USSR you have a laboratory that has existed for more than twenty years, and you think that Marx should know more than you about socialism. You see, Marx didn't predict this or that. . . . You need to work with your own heads and not string together quotations. There are new facts and new combinations of forces. Be so kind as to work with your heads."[9] Stalin's disdain for the blind application of Marx and Engels did not mean that their categories of economic analysis should be abandoned. To the contrary, during the transition to communism economic ideas associated with capitalism were evidently still valid. In the textbook Leont'ev presumed that the "law of value," Marx's key to understanding the capitalists' exploitation of workers, had been overcome in the USSR. Stalin disagreed, noting that the category of the cost of production, which depends on the law of value, was necessary for commodity distribution and for setting prices. The existence of "illegal markets" and kolkhoz markets meant that market prices still existed and that the government was not in total control of prices. The USSR was not able to distribute entirely according to need, and therefore the law of value was still used. Stalin gave the example of a bread shortage in the newly Sovietized Baltic region in 1940. Due to bad crops, the market price of cereals rose quickly. The government sent a shipment of cereals to the region, and the prices dropped. This was an example of prices still being dictated by the law of value. But until the USSR had ample reserves of all commodities, the government would be unable to perform similar procedures each time a crisis arose. As long as kolkhoz and other markets existed, prices could not be totally controlled, and the law of value would remain in force. Nonetheless, Stalin distinguished between the law of value under capitalism, where it "exists spontaneously, brings destruction, and requires enormous sacrifices," and the law of value in the USSR, where its "character changes," taking on "new content and new form." In the USSR, he declared, by consciously using the law of value to help distribute goods, the government minimized the need for sacrifice.[10]

Stalin's attention to the content of the textbook was matched by his concern about the overall style and tone Leont'ev employed. Stalin wanted the book to be based in "science." Instead he found that "the language of propaganda leaflets and posters appears out of nowhere in the textbook. This will not do. An economist should study facts, but suddenly we get 'Trotsky-Bukharinist traitor . . .' Why talk about the fact that the court established this or that? Where is the economics in this? The propaganda should be tossed out. Political economy is serious work." No other Soviet leader could have confidently implied that propaganda somehow distracted from "serious work." Stalin, of course, remained the judge of seriousness and propaganda. But for him the distinctions were clear; science was the standard for political economy. As he told the audience in his Kremlin office, "In science we appeal to the mind. But here [in the textbook], the appeal is to the gut or to something else. This ruins the work." Given these standards, Stalin claimed to have little patience for exaggerations of Soviet accomplishments. "You embellish our reality too much. By no means have we become as pure as we would like. Criticism of our practical work is needed." Later in the meeting, he declared, "You need to write in such a way that it doesn't come across that everything is bad for them and everything is good for us."[11]

The January 1941 meeting clarified that Marx's categories were applicable to the USSR, but Stalin left the task of defining and discovering other "laws of socialist economics" to a commission of authors headed by Leont'ev. Stalin gave no concrete clues as to how the book could be both Party-minded and based in "real laws" while describing "socialist realism." It also remained unclear exactly what economic laws existed and whether they in any way restricted Stalin, the state, and the Party. Did political decisions or objective economic laws lead to the development of socialism? In March 1941, Leont'ev wrote to Andrei Zhdanov that attempts to complete the latest draft of the book had become bogged down. He reported that his "efforts to pin down the commission's broad opinions about the more difficult and debated questions resulted in little of value since nearly all of these questions eluded consensus." A few days later, however, he added that Stalin's corrections had been integrated into the sections on capitalist and socialist modes of production.[12]

Despite Stalin's effort, an approved version of the textbook did not make it out before the beginning of the Second World War. During the war, work on the textbook all but stopped, although from 1943 to 1945 *Under the Banner of Marxism*, the main theoretical journal, and *Bolshevik*, the organ of the Central Committee, attempted to elaborate on some of Stalin's points.[13] Meanwhile, the war itself and its aftermath introduced myriad new economic phenomena that had to be explained. How had socialism functioned during the war, and was the 1930s policy of preparing for war at the cost of consumer goods now open to reevaluation? What did the war tell Soviet economists about capitalism? Had the experience of war altered concepts such as the inevitability of conflict between imperialist countries? How did the "people's democracies" forming in Europe affect the building of socialism? Were there many paths to socialism? What was the USSR's role in aiding the new socialist countries? The USSR was no longer "encircled by capitalist powers." Did this mean that there could be an easing of the pressure to produce the means of production? In short, the task of writing a textbook on political economy, already difficult in the 1930s, was now made infinitely more so by unforeseen developments.

Within days of the end of the war in Europe, Leont'ev once again wrote to Andrei Zhdanov stating that work on the political economy textbook could now be completed, even though "the draft of the textbook edited by the Central Committee commission on the eve of the war clearly needs some reworking."[14] Zhdanov, evidently confident that the book could be finished quickly, forbade publication of other material on the political economy of socialism until the book's release. By late 1945, however, the book had not been published, and teachers were concerned about what ideas they were supposed to present in class. When the director of the Central Committee's Higher Party School pleaded with Malenkov for permission to print copies of an old pamphlet on political economy for use in the classroom, he was told to sit tight and to wait for the *Short Course on Political Economy*, which was "almost done."[15]

Finally, a draft of *Political Economy: A Short Course* was printed in limited numbers in 1946.[16] In April 1947 the Politburo created a three-person commission of Andrei Zhdanov, Nikolai Voznesenskii, and Leont'ev to edit the draft and see it through to completion in three months.[17] The commission called in dozens of scholars to help on the project. They also sent copies of the draft to economists and teachers asking for comments on where additional work was required. A year later, two versions of the textbook had reached some degree of completion. The first consisted of new chapters written by Leont'ev in response to Politburo instructions and revised older chapters based on the reviews he had received from his colleagues. The second version of the textbook contained new information on precapitalist political economy and had been written under the directorship of Ostrovitianov. Both versions had identical chapters on socialism. Zhdanov, Voznesesnkii, and Leont'ev reported to Stalin in April 1948 that the next step was to unite both versions into a uniform textbook covering political economy from primitive communism through advanced socialism. In the summer of 1948, however, Zhdanov died, leaving the book without a clear political patron.[18]

Figure 13. Politburo members, 1945. Shown at Stalin's right (*from left to right*) are Nikolai Voznesenskii, Lavrenty Beria, and Georgii Malenkov. Courtesy of RGAKFD.

The demand for the textbook had not let up, however. In the fall of 1948 the minister of education, Kaftanov, wrote to Malenkov complaining that though political economy was required throughout the educational system, there was still no textbook on the subject. This proved to be especially problematic for teaching the political economy of socialism. When asked to review the matter, members of Agitprop recommended that the latest draft be published with a small circulation and a note stating that it had been approved by Agitprop or the Ministry of Education but not the Central Committee. This would have allowed economists to teach the subject before the Party had passed final judgment. Such a compromise was evidently too risky: the Secretariat looked into the matter in late 1948 and decided to delay any further decisions.[19]

Leont'ev and the others who were brought in to help with the textbook faced a number of problems, which could not be solved without Stalin's further guidance. They could not rely solely on what he had said in 1941. As world politics and Soviet politics changed, so too did economic theory. For instance, reviewers ridiculed the 1947 version of the textbook because it had not provided the theoretical justification for the monetary reform of 1947, a policy that had been implemented in the period between the book's completion and the time it was under review. The task of writing the textbook was further complicated by the fact that the field of political economy had been rocked by a series of controversies. From 1947 to 1949, academic meetings, Central Committee memos, and articles in popular and scholarly publications denounced Evgeny Varga, the USSR's premier expert on the political economy of capitalism, as a "revisionist," "anti-Leninist" "cosmopolitan" who had produced "politically undependable" work. He had challenged orthodox ideas by suggesting, among other things, that bourgeois states were playing an increasing role in regulating their economies and that the Eastern European democracies were "state capitalist" rather than socialist.[20] With Varga in retreat and the Party dismantling his Institute for World Economics and World Politics, Leont'ev had few people to turn to for help with the textbook's sections on modern capitalism.

The section on the political economy of socialism was proving just as difficult. In 1947 it appeared as though Voznesenskii, as a Politburo member and Gosplan director, might come to the rescue. His book, *The War Economy of the USSR during the Great Patriotic War*, argued that, through careful planning and the use of the "transformed law of value," the state and Party had been able to control costs, production, and distribution. He outlined an approach to the political economy of socialism that was governed by economic laws but also beholden to a centralized system of planning. The "spontaneity" that characterized the economy in the early 1930s had given way to carefully organized and authoritative planning based on objective economic laws. The book received extremely positive reviews in all the major Soviet journals. More important, it seemed to offer Leon'tev a model of how socialism worked in theory and in practice.

While it is not certain, praise for Voznesenskii's book may have upset Stalin, who believed that he was the ultimate authority on the Soviet economy. In any case, as part of Malenkov and Beria's efforts to remove Zhdanov's protégés from power, Voznesenskii lost favor with Stalin, was fired, and then was arrested. In 1950 he was shot without a trial.[21] Voznesenskii's fall from power threw Soviet economic theory into turmoil. It was unclear whether his views about the political economy of socialism were completely erroneous. Gosplan, which he had controlled, was responsible for overseeing organizational and scholarly work at the Institute of Economics. Was all the work conducted under Gosplan now supposed to be refuted? What about Voznesenskii's theoretical statements? The chances of producing an acceptable political economy textbook decreased as a result of Zhdanov's death and Voznesesnkii's arrest, but the need for the textbook became all the more urgent.

The troubles in the field quickly spread to major economics institutes and publications. In 1949 and 1950 the Central Committee established a series of commissions to look into a wide range of problems afflicting research, publication, and education in the USSR. The field of economics was particularly hard hit by the "anticosmopolitan" campaign. A large number of Soviet economists, especially those who had studied capitalist countries at Varga's institute, were "ethnic" Jews, meaning they were identified as Jewish in their passports regardless of their religious beliefs. Varga and I. A. Trakhtenberg—each

of whom had met with Stalin in the past—came under intense attack in early
1949 as "cosmopolitans." Leont'ev, also a Jew, was accused of supporting Varga's views, essentially because he had not joined the chorus of economists
who had criticized Varga's work over the previous few months. While these
prominent economists avoided severe punishment, others were not so fortunate. One report to the presidium of the Academy of Sciences in late March
1949, when the worst of the campaign had already passed, stated that around
fifty people had been fired from the Institute of Economics. Ominously, the
report suggested that "this work, clearly, must continue, and a few more people
need to be let go in order to rejuvenate the cadres, cure the unhealthy atmosphere, and silence once and for all the cosmopolitan spokesmen." Ostrovitianov, as the director of the institute, gave a detailed list of who should be kept
and who should be fired from among those charged with being "cosmopolitans." Varga was spared a serious reprimand. The Agitprop leader Shepilov had
evidently told Ostrovitianov that "there is no need to alienate him [Varga]; if
he voluntarily [admits his mistakes], we need to help him." Like Varga, almost
all the potential victims were Jewish, and the names of those Ostrovitianov
suggested letting go (Blumen, Roitburg, Lemnin, etc.) show that the campaign
took the form of thinly veiled anti-Semitism.[22]

As the head of the Institute of Economics and editor of the journal *Questions of Economics*, Ostrovitianov also faced harsh criticism, although not for
cosmopolitanism. The Party commissions looking into economics consistently
accused the institute of "serious shortcomings" including the failure to produce
"concrete, objective scientific research." Part of the problem was that when
Voznesenskii was still alive and in control of Gosplan, the institute and journals had used his work as the basis of their scholarship and teaching. Ostrovitianov responded by invoking the "anticosmopolitan" campaign on his own
behalf. He reported to the Orgburo that there remained a "dangerous monopoly of old cadres, especially those working on questions of modern capitalism,
who prevent the advancement of talented younger people and who present a
danger of a relapse of reformist mistakes at the institute." Who were these
monopolists? He reported that "Of the 83 senior researchers . . . 72 were members of the Party, but 21 of those had once been in the [Jewish] 'Bund' and
other parties and had deviated from the Party line and Party rulings." Of
the 83, only 44 were Russian, whereas 34 were Jewish and 5 were of "other
nationalities." Furthermore, Jews dominated the ranks of the most qualified
economists. Of the 34 academicians, corresponding members, and doctorates
at the institute there were 20 Jews, 12 Russians, and 2 members of other
nationalities.[23] Some Jews responded to anti-Semitism by using Russian pseud-
onyms when they published their scholarly work. But Yuri Zhdanov reported
to the ideological chief Suslov that the practice was becoming systematic and
troubling, because it "weakened authors' sense of responsibility for their work
and at the same time creates the incorrect impression that their work can only

be published under Russian last names."[24] In a profession dominated at its highest levels by Jews, the fact that Ostrovitianov was Russian could only have helped his cause in the Party. Although the Central Committee considered removing him from his post a number of times, it always backed down.

Problems ran deeper than the institute's leadership. Economists of all stripes avoided addressing questions on the political economy of socialism, choosing instead to wait for the publication of the political economy textbook. A Central Committee report written by Yuri Zhdanov concluded that the opposite was needed: economists should produce work that could help with the drafting of the textbook.[25] The crisis in Soviet political economy had entered a vicious circle. Without scholarly articles to rely on, editors of the textbook could not possibly address all the problems in the field of political economy. Without the textbook on political economy, scholars were fearful of writing articles that might be refuted when the definitive work came out.

The writing and editing of the political economy textbook took place against a backdrop of crisis in the field. Theoretical disputes overlapped with political battles that in turn reflected institutional problems. Added to the mix was an increase in semiofficial anti-Semitism, which was bound to have adverse effects on a field in which Jews played a prominent role. Without a definitive political economy textbook, scholars were reluctant to forge new economic ideas or commit themselves to all but the most superficial positions. This only increased the pressure on the writers and editors of the textbook.

In the late 1940s the Party secretaries Malenkov and Suslov and the Agitprop chief Shepilov did not fill the void left by Zhdanov and Voznesenskii. Instead of showing interest in the fate of the political economy textbook, they concentrated their efforts on the Institute of Economics. They must have concluded that the textbook was an ill-fated project. But Leont'ev, Ostrovitianov, and others continued to work on the textbook, attempting to keep pace with debates in the field. By late 1949 they had completed two overlapping versions of the textbook and sent both to Stalin, who remained determined to steer the book through to completion. In December Stalin discussed the two versions with Leont'ev by telephone and read up on various conflicts among the authors.[26] On February 22, 1950, Stalin summoned Malenkov, Ostrovitianov, Leont'ev, and the philosopher Pavel Iudin to his Kremlin office at 11:15 p.m. to discuss the drafts he had received.[27] Stalin decided that Leont'ev's version should be used as the foundation for a single textbook but suggested that the section on American imperialism be strengthened based on articles from *Bolshevik* and *Questions of Economics*.[28]

International politics had clearly influenced Stalin's approach to the textbook. In late 1949 and early 1950 Stalin held meetings in Moscow with Mao

Zedong and Zhou Enlai to discuss a Sino-Soviet treaty. On February 14, leaders from the USSR and the People's Republic of China signed the Treaty of
Friendship, Alliance, and Mutual Assurance. Not surprisingly, those parts of
the political economy textbook that addressed Chinese communism had
caught Stalin's immediate attention. He spent most of the February 22 meeting with Malenkov, Ostrovitianov, Leont'ev, and Iudin discussing the distinction between the People's Republic of China and the people's democracies of
Central and Eastern Europe. Rather than placing the two groups in the same
category, Stalin informed the two economists that China still lagged behind
the USSR and even Eastern European countries on the road to socialism. He
explained that China had a dictatorship of the proletariat and peasants, that
its revolution had included Chinese bourgeoisie who were rebelling against
foreigners, and that feudal relations, to a great extent, remained in force in
China. In contrast, the European people's democracies had dictatorships of
the proletariat, nationalization of industry, had achieved socialism in the
cities, and were now working on bringing socialism to the countryside. China's
people's republic was "still in the first phase of development." Stalin suggested
that the confusion about this matter stemmed from the fact that "our cadres
do not have strong training in economics." But he did not provide practical
advice about how the textbook writers should deal with contrasting roads to
socialism. Was the textbook about the USSR's particular road to communism
or was it about explaining various contexts in which socialism may develop?
Were its writers supposed to encourage many paths? Without answering these
questions, Stalin ordered the other participants in the meeting to form a commission to "rework the draft in one month."[29]

The very fact that Stalin met with the textbook writers in the wake of his
intense negotiations with the Chinese suggests the significance he assigned
the book. The economists must have felt privileged. Party secretary Malenkov
came away from the meeting sensing he needed to take a more active role in
bringing about the book's completion. The bureaucracy responded. Less than
two weeks later, Minister of Education Kaftanov sent Malenkov and Suslov a
draft resolution emphasizing the book's central role for higher education and
outlining a plan for completing the project. He repeated a concern he had
raised a few years before: despite the fact that every institution of higher and
technical education in the country had a course on political economy, no
textbook on the subject existed. Kaftanov proposed that the Central Committee appoint an "authors' collective" to write the textbook, and he suggested
nine authors who would work under the editorial leadership of Ostrovitianov
and Leont'ev. He also requested that the group of authors be given from three
to five months off from their other professional responsibilities in order to
finish the book in just under one year.[30] The Secretariat passed a series of
decisions giving Iudin, Ostrovitianov, and Leont'ev leaves of absence from

their primary jobs in order to "fulfill a Central Committee assignment."[31] In March and April 1950, Malenkov and other members of the Secretariat monitored economic scholarship closely.[32] Stalin's involvement had placed the textbook back on the Party's main agenda after it had been an inconsistent priority for over a decade.

By April the commission had submitted a single textbook to Stalin, and within days he convened a meeting at his Kremlin office that included the participation of six members of the Politburo and deputy premiers of the Council of Ministers—Lavrenty Beria, Nikolai Bulganin, Lazar Kaganovich, Anastas Mikoian, Malenkov, and Molotov. Shepilov, now a Central Committee inspector, also joined Leont'ev, Ostrovitianov, and Iudin for the meeting, which once again took place late at night.[33] Stalin began with criticisms of the section on precapitalist and capitalist modes of production, leaving "socialism for another time." Just because Stalin had approved Leont'ev's version of the textbook as a starting point in February did not mean that "major revisions" were not needed. The book required "serious corrections" in both tone and substance. He accused the authors of displaying a "complete misunderstanding of Marxism." The book spent too little time on the machine period of capitalism, which Stalin believed was responsible for the transformation of society. In this respect, Stalin called on the authors to follow Marx's lead, since *Capital* devoted 110 pages to the machine period and only 28 pages to the manufacturing period.[34]

Marx and Engels, however, could not provide answers to all the questions addressed in the textbook. Noting that "a lot of time has passed since Marx," Stalin called for a fresh assessment of wages during the period of monopoly capitalism. Recognizing that the textbook would be a weapon in the Cold War, he told the authors that "we are in a struggle with capitalism right now on the basis of wages." He implored them to "take real facts from contemporary life" and "use concrete facts to show the situation with real wages here," and he emphasized how the textbook should reveal that wages in the West did not provide workers with a living wage, despite the propaganda of capitalist countries.

Stalin's complaints ranged widely. He demanded less space in the textbook for primitive accumulation and more criticism of the idealist views of the "utopianists." Admitting that in *Capital* Marx concentrated on industry, Stalin wanted greater attention given to agriculture in the descriptions of both capitalism and socialism. "We have a different task before us [than Marx did]," he informed the authors, adding, "we know what kind of weight and meaning agriculture has in the economy."[35]

Stalin was also critical of the book's "many babbling, empty, and unnecessary words" and the "poorly developed" literary style. He insisted on the book's reorganization, a more refined definition of political economy, and the use of the historical, as opposed to the analytical, method in structuring the book. Finally he suggested that the contemporary lack of knowledge of Marxist economic theory was potentially devastating to the welfare of the state and Party. He commented that "it is bad that there are no disagreements in the commission and that there are no arguments over theoretical questions," and added, "I mean, you are involved in a historic undertaking. Everyone will read this textbook. Soviet power has been around for 33 years and we don't have a book on political economy. Everyone is waiting." In lamenting the lack of arguments among the economists, Stalin was evidently reminiscing about an older group of Bolsheviks who had originally been his peers but whom his regime had all but eliminated. He explained the importance of the textbook to Party leaders and economists by putting the issue in historical perspective:

> The first, older generation of Bolsheviks was well grounded theoretically. We memorized *Capital*, summarized, argued, and tested one another. This was our strength. This helped us a lot. The second generation was less prepared. People were busy with practical work and construction. They studied Marxism through brochures.
>
> The third generation has been raised on pamphlets and newspaper articles. They don't have a deep understanding of Marxism. They must be given food that is easily digestible. The majority of them were raised on quotations, not the study of Marx and Lenin. If things continue this way, people might degenerate. People may decide they don't need *Capital* when we are building socialism. This threatens degradation. This will mean death. In order to avoid this even in part, it is necessary to raise the level of economic understanding.[36]

Stalin's dramatic—even prophetic—warning reveals that he recognized a growing discrepancy between the Party's official ideology and Soviet practice. Stalin (and hence everybody else) was in a bind. Ideology and reality were supposed to be one and the same, but they appeared to be diverging. In 1941 he had told the economists to forget the clichés and verbiage culled from Marx and Engels and turn their attention to studying Soviet reality. Now he added that the only way to understand reality was to study and argue about the classics. Invigorating the work of the "third generation" was one of the central goals of the new textbook. As Stalin pointed out, the stakes for the regime could not have been higher. But determining where Marx's work remained relevant and where it needed to be revised was too dangerous for anyone but Stalin himself to attempt.

Toward the end of the meeting, Stalin informed the authors that "when the textbook is finished it will be placed before the judgment of public opinion."[37] This may have been an early indication that the textbook would be the subject of a large discussion along the lines of those that had taken place in other fields of scholarship. Discussion, Stalin implied, would help "raise the level of economic understanding."

A little over a month later Stalin again met with the authors of the text-book at his Kremlin office. Stalin offered detailed criticism of the section of the book on feudalism, the use of Marx and Lenin, and the overall style of the book. Noting that "very little is said about Russia" in the description of feudalism, Stalin wanted more examples from Russian history right up through the peasant reforms and the abolition of serfdom. He also continued to resist the rigid use of Marx, reminding the authors that "Marx wrote in the 1860s and since then technology has moved forward."[38] A postwar Soviet textbook had to reflect the Russocentrism of the period even when doing so meant deflecting attention away from Marx's analysis of economic history.

While conceding that "writing a textbook is not a simple affair," Stalin urged the authors to "ponder history more" and not to become complacent in the presentation of their ideas. At times, Stalin said, the textbook "used bizarre propaganda and popularizing language," making it seem "like some grandfather telling fairy tales." Again, he reminded the authors of the high expectations he had of their work: "The textbook is intended for millions of people. It will be read and studied not only here, but all over the world. It will be read by Americans and Chinese and it will be studied in all countries. You need to keep in mind a more qualified audience."[39] Stalin recognized that revising the textbook would take a long time and, noting that this was "very serious work," suggested that the commission be given at least until the end of 1950 and possibly into 1951 to complete their work. The result promised to be significant: Stalin informed the authors that he intended to put "Approved by the Central Committee of the Communist Party" on the title page of the finished product.[40]

Stalin's personal involvement sent a signal to everyone involved that the textbook was more than just another piece of Soviet scholarship. But such close scrutiny from Stalin hardly made the task of writing the book any easier. After all, on the political economy of socialism—the subject that was most rife with controversy and most in need of official formulation—Stalin had barely said a thing. Economists were left with the task of describing socialist theory and the "real existing socialism" that was around them.

No other work of scholarship in the postwar period, and perhaps in all of Soviet history, was subject to such persistent attention from all levels of the Central Committee. Completing the textbook became a group project directly involving all of the nation's prominent economists and many of its most pow-erful people. Immediately following the Kremlin meetings the "authors' col-lective" (now consisting of Leont'ev, Ostrovitianov, Shepilov, Laptev, Iudin, and Pashkov) began working on producing an acceptable version of the book. Stalin and Malenkov closely oversaw their progress, occasionally offering feed-back on particular sections. The first order of business was to establish a plan for revising the existing manuscript. On June 9 Shepilov sent Malenkov the authors' plan for the section on primitive communism, slave owning, and feudalism, about which Stalin had been most critical. As they informed Stalin less than a week later, they had revised the first four chapters of what now had become a single section on "precapitalist" modes of production. In keeping with Stalin's instructions, other changes were made in the book's organization.[41] In their letter to Stalin, the authors illustrated how they were following his specific instructions. Stalin read the drafted chapters, cutting whole paragraphs, adding phrases, and making a number of simple editorial corrections.[42]

In the summer of 1950 Shepilov sent Malenkov an updated work schedule for the book's production, identifying those responsible for writing and editing each specific chapter and target dates for the chapters' completion. For the next three months, Shepilov repeatedly reported back to Malenkov on the authors' progress and plans. The writers, who essentially stopped working at their regular jobs, were given the use of a large dacha outside Moscow as well as a special car and driver.[43]

By January 1951 the authors had sent Stalin and Malenkov twenty-five of the book's thirty-four chapters and reported that they could finish the rest by March 30. Around the same time, the Secretariat gave the authors special permission to take three more months off from other work in order to complete the manuscript. Finally, they finished the final nine chapters on the socialist mode of production, receiving the last two from the printers on April 25, 1951. That same day, the six authors reported to Stalin and Malenkov that they had completed the Central Committee's assignment. They enumerated the ways in which the textbook corresponded with Stalin's orders: they had used the historical method; emphasized the machine period of capitalism; concentrated on wages and workers' conditions in both capitalism and socialism; and added information on agrarian relations during capitalism. They also reported that "the bulk of the text of the textbook has been rewritten from scratch and the text taken from the earlier versions underwent serious revision."[44] This time Stalin decided not to edit the textbook himself, opting instead to sponsor a large discussion at the Central Committee.

In early July 1951 the Central Committee distributed nearly 250 copies of the latest draft of the textbook—stamped with the Central Committee's seal—to economists, teachers, and Party and industrial leaders. Significantly, even many of those who were quite critical of Leont'ev and Ostrovitianov were included. A cover letter suggested that the book had broad significance for Soviet citizens and for the Cold War struggle:

> The Central Committee of the Communist Party (Bolsheviks) attaches great importance to the creation of a full-fledged political economy course. Such a course is necessary to improve the Marxist-Leninist education of the Soviet intelligentsia

and to help our cadres efficiently solve practical questions relating to the further development of the socialist economy. It is necessary not only for our country but also for all the people's democracies and for the communist parties in capitalist and colonial countries.[45]

The letter also made it clear that the discussion was not organized simply to endorse the book:

Keeping in mind that the prepared textbook is in need of serious improvement, the Central Committee believes it necessary to conduct an open discussion about the enclosed textbook so that economists' useful critical comments and suggestions can be taken into account in the final editing of the course. In sending you a copy of the draft of the textbook, the Central Committee invites you to participate in the discussion, which will take place in Moscow in the middle of September of this year.[46]

In late October 1951 Suslov and Yuri Zhdanov wrote to Stalin and Malenkov outlining the ground rules for the discussion. The meeting would take place beginning on October 26 in either the Sverdlov Hall in the Kremlin or in the conference hall of the Central Committee apparatus "where the philosophy discussion took place." The memo left unresolved the questions of who would open the meeting and who would preside over it but noted that the 247 people (182 from Moscow and 65 from other towns throughout the USSR) who had received the draft of the textbook would participate. Suslov and Zhdanov anticipated that somewhere between forty and forty-five people, including the book's authors, would address the meeting, which was scheduled for four hours each day for six to seven days. Those unable to speak would be able to contribute their ideas in written form. Though no information about the discussion was to appear in the press before its conclusion, Suslov and Zhdanov did anticipate publishing a stenographic record as a separate book. All recipients of the draft of the textbook were expected to return their copies to the Central Committee after the meeting's conclusion. Suslov's and Zhdanov's memo on the discussion's organizational structure and a draft Central Committee resolution outlining their basic points were also sent to other Politburo members including Beria, Bulganin, and Kaganovich. On October 25, the Politburo approved the plan but delayed the start of the discussions for two weeks.[47] In advance of the meeting Zhdanov and an assistant from Agitprop sent Malenkov a long memo outlining responses to the textbook from seventeen different economists working at the Institute of Economics and elsewhere in Moscow. While one or two of the selected economists evidently believed that a "discussion was not necessary," most outlined specific problems with the book, giving Malenkov and the Central Committee an adequate idea of the types of arguments and suggestions that were likely to arise in the discussion.[48]

The meeting finally began on November 10, 1951, and took place in the Central Committee building with Malenkov presiding and Suslov acting as his deputy.[49] In his opening remarks, Malenkov repeated that the book was in need of serious criticism but also suggested that the authors had done good, positive work and that the draft contained much useful material. He called for a full and free discussion of the book's merits and weaknesses and emphasized that the authors would listen to all critical comments. The Central Committee set aside two weeks for the discussion, with meetings lasting from 11:00 a.m. to 6:00 p.m. every day except Sundays. The participants had a one-hour lunch break as well. When asked what the participants were supposed to do about their lectures and regular jobs while the meeting was under way, Malenkov responded that all other responsibilities should be rescheduled and that the Central Committee would take care of any problems that might result. As the schedule was set up, there was time for about six to eight speakers a day.[50]

One of the first issues addressed by the participants was whether the textbook was useful in any way or whether it should simply be tossed out. Although everyone complained about some aspect of the book, most saw their task as providing constructive criticism in order to improve the existing text. As one participant pointed out, the textbook could not be dismissed as useless, nor was it in good enough shape to publish as it was. Participants knew that previous discussions had resulted in clear scholarly "winners" and "losers." Since Stalin had been directly involved in the content of the textbook, the potential consequences this time must have seemed particularly significant. Some participants looked at Malenkov or Suslov in the chairman's seat and attempted to say what they suspected the Central Committee wanted to hear. If that meant the textbook should be accepted as a basis for discussion and also criticized, so be it. Others spoke their mind, letting the Central Committee see "what [they were] thinking," even at the risk of future ridicule but also, perhaps, in the hope of influencing the outcome.[51] Of all of the scholarly discussions of the postwar period, this one was most free ranging, with few completely uncontested arguments. Though a few of the speakers on the first days argued that the textbook marked a step backward for economic scholarship, the vast majority disagreed and, despite some criticism, argued that the book provided a valuable starting point. As a result, there were no clearly delineated sides in this discussion. Few positions were dismissed outright. In this case, there were no equivalents of Lysenko, Bykov, or Andrei Zhdanov to rally around, nor any "Mendelist-Morganist," "anti-Soviet," or "bourgeois" lines to denounce. Instead, the discussion centered on the problems of the textbook specifically, the reasons for stagnation in Soviet economics more broadly, and the tensions between Marxist-Leninist doctrine and Soviet reality in general.

Disagreements abounded about what a Soviet political economy textbook was supposed to accomplish. Some thought that it should be a definitive work: "not a single problem and not a single thesis should raise a shadow of a doubt that the textbook will give a clear and final answer to all questions."[52] This rigid understanding of the textbook's purpose went hand in hand with the notion that Soviet political economists should possess all the answers to all the difficult questions facing them. The field required its practitioners to define and defend orthodoxy as well as develop new economic theories. Along these lines, a researcher at the Institute for International Relations complained that the authors of the textbook, for fear of making a mistake, ignored certain questions that remained contentious among economists. He implored the authors to write about every subject and allow criticism of their work to help them along. A definitive textbook could not equivocate.[53] Varga seemed to have an impossibly high standard for the textbook, arguing that "the book still does not have what is needed in a good textbook: that the meaning of each phrase is clear, that each sentence is exact, that no statements contain discrepancies and double meanings."[54] Who could write with such arch authority? The goal, it seems, was to write the Soviet New Testament, an updated and improved version of *Capital*.

Other scholars took the opposite approach, arguing that the authors had to limit the textbook's scope and the number of economic problems it could resolve. Academician T. S. Khachaturov pointed out that questions being disputed by specialists could not be solved in a general textbook: "You would like for the authors of the textbook to resolve dozens of raw theoretical questions for all Soviet economists and then to include their solutions in the textbook. We should not count on that happening."[55] The deputy director of the Institute of Economics, V. P. D"iachenko, agreed. There was too little space in the textbook to cover all questions of political economy. The authors were better off limiting the scope of the book to the most important issues and illustrating them well.[56] Another scholar pointed out that "the political economy textbook is not a reference book with prepared answers to every possible question of the social-political life of the past and present." He noted that some of the subjects in the textbook were better left to histories of specific economic categories.[57] Trakhtenberg concurred, reminding the audience that "we are not talking about a multivolume monograph where almost every problem of economic theory is addressed." To the contrary, the textbook was written as a popular, introductory course.[58] Some questions were inevitably going to be left open ended. Those scholars looking for a definitive textbook that would address and answer all the theoretical disputes of the period benefited from the fact that the textbook itself had set out to be both chronologically and theoretically all-encompassing. This meant that any subject touched on in the textbook was fair game for the discussion participants.

Disputes about the content of the book evolved into disputes about the state of the field in general. Each theoretical point in the textbook relied on the views of countless specialists. Debates were particularly heated about the economic foundations of socialism, where specific subjects often led to more fundamental questions about the relationship of scholarship to the Party and its doctrines. The socialist mode of production was by far the most contested section of the book. While the authors could lean on Marx, Engels, and Lenin to describe capitalism and imperialism, there were fewer canonical texts for socialism. The years of scholarly stagnation in the field did not help the situation. The authors of the textbook could depend on their conversations with Stalin as a reliable source, though in the most recent meetings Stalin had not addressed economic problems of socialism. Paradoxically, the textbook's emphasis on the existence of a "transformed" law of value under socialism, a thesis the authors had received directly from Stalin in 1941, became the most controversial subject in the discussion. The law of value raised questions about the functioning of economic laws of socialism and the relation of the state and Party to those laws. The economists had been called on to define the nature of the Soviet system while remaining under doctrinal constraints and without clear guidance from Stalin.

One group of economists argued that the policies of the Soviet state and Party created and defined all economic laws. Inasmuch as the Soviet state controlled both the politics and economics of society, then it was part of both the base and the superstructure. The state itself should be treated as an economic category.[59] On the opposite extreme, Ia. F. Mikolenko, a professor at the Moscow Institute of International Relations, argued that economic laws of socialism unfolded spontaneously—that is, independently of human will— and reflected objective necessities. In his view, economic laws could not be created or destroyed, and the actions of the state and Party were subordinate to these economic laws. For Mikolenko, the question was a matter of common sense: "A law that does not exist at all times, and is not uniform and is somehow dependent on human will or understanding, is logically meaningless. The very definition of a 'law' is that it exists at all times, is consistent, and unavoidable."[60] His ideas were attacked by subsequent speakers, who sought a middle ground between defining economic laws in totally "objective" terms and ceding to those who wanted to devolve every economic law to a matter of the will of the people and the Party.

While most participants could accept "planning" as an economic law of socialism, there was plenty of debate about the precise meaning of the "transformed law of value." One group of economists argued that commodities and value under socialism were completely different categories than under capitalism. In the USSR, they resulted from society's need to distribute according to labor. For instance, laborers working for the state, cooperatives, and private kolkhozes all had to be able to exchange products with one another. In theory, by using the transformed law of value, the state was able to assign value to work and commodities without the exploitation that came with added value under capitalism.

Academician S. G. Strumilin saw the law of value as an eternal law that regulated exchange between sectors of society and was also a tool for regulating industrial production. He believed that even under communism, the law of value would continue to serve this second function.[61] Perhaps the most ridiculed explanation for the law of value under socialism came from M. N. Meiman, who argued that the law of value existed simply because the USSR was surrounded by capitalist countries. The maintenance of commodities, value, and prices was simply a way for the Soviet Union to hide the details of the economy from the outside world.[62]

Given the disparate views about the law of value in the USSR, it was no wonder the textbook had been unable to offer a durable definition. As one sympathetic speaker noted,

> Each comrade speaking from the lectern about the law of value felt it necessary to develop his own conception of the law under socialism. More than 20 different conceptions have been put forth here, which is quite a bit more than is needed. (laughter) Nine different workers from the Institute of Economics have formulated 9 different conceptions about one and the same law. . . . One of these comrades who spoke about the law of value said of his own talk: "Although I developed my point of view about the law of value, I'm still not sure exactly what that point of view is." (laughter)[63]

As the discussion evolved, a consensus formed around the idea that commodities and value were historical categories inherited from the capitalist mode of production and fundamentally transformed by economic planning and the socialist control of the means of production. But general statements about the law of value's benign existence did not substitute for detailed analysis, and it was left to the authors of the textbook to work out a tenable description of the law of value in practice.

Many participants seemed to recognize that the existence of economic laws provided economists with an independent professional identity. If economic laws existed, experts had to study them. However, if economic laws were completely subordinated to the wishes of the state and Party, then economists were left in the position of trying to justify political decisions in economic terms. Certainly in Stalin's time, there was little opportunity for economists to mount a challenge to political doctrines based on their analysis of real economic conditions. But the existence of economic laws would leave open that possibility for a later time, when political authorities would cede more room to scholars.

If economic laws existed, how were economists supposed to discover them? Using the historical method, as Stalin termed it, allowed the authors of the textbook to show economic laws unfolding as history progressed. This raised some serious questions about the general meaning of those laws and their applicability beyond specific historical moments. Until the 1940s Soviet econ-

omists relied solely on the history of the USSR to distill all the general laws
of the political economy of socialism. But the appearance of other socialist
countries weakened the value of such an approach. If the authors gave too
many specifics about Russian historical development, it might raise questions
about socialism elsewhere. If they spent too little time describing Russian
and Soviet history, they could be accused of abstractions and abandoning the
historical method. Participants in the discussion were quick to point out where
they believed there needed to be more emphasis on capitalism in Russia and
more tightly argued passages on the relation of Soviet history to the general
understanding of socialism as it would evolve elsewhere.

The Russocentrism of the postwar period complicated the question of how
much emphasis to give Russia's pre-Revolutionary history. Previous scientific
discussions had indicated that Russia's past was supposed to be emphasized.
Participants in the political economy discussion knew this and called for more
attention to the development of capitalism in Russia in order to explain why
socialism had occurred there and not elsewhere. This would also give Russian
readers "familiar material" that would be easier for them to understand. But
emphasizing tsarist Russia's imperialism and capitalist exploitation hardly fit
with the pro-Russian rhetoric of the period. So rather than emphasizing Russia's economic history, other participants suggested the authors give a fuller
treatment of Russian revolutionaries from the Decembrists through Chernyshevsky. One scholar felt that the textbook gave short shrift to the great discoveries of Russian scientists in the nineteenth century. Including a section
on Russian contributions to science and technology would make the textbook
patriotic without overly emphasizing the negative aspects of Russia's past.[64]
Academician Strumilin saw theoretical problems with using Russian economic history to teach about capitalism in general: "If our manufacturing
period lasted until the beginning of 1890, the period of imperialism in Russia
began no later than 1899–1903, which means that the period for the premo-
nopoly machine period would be less than ten years, and even then without
a single crisis of production. Capitalism without crises is not the best basis for
building economic theories."[65] While using Russia as the primary example
for explaining the development and evolution of capitalism hardly seemed
reasonable, the Soviet Union provided economists with their only test case
for developing laws of socialism. The very first speaker at the discussion
pointed out that it was not clear whether industrialization was a law of socialism only in the USSR or whether it should be more broadly applied. Echoing
this concern, another speaker believed that the textbook failed to show the
"world-historical significance" of Stalin's theory of collectivization. Collectivization of agriculture was a "law of socialist development"; that is, it was a
necessary element in the path to building socialism.[66]

So long as the USSR was the only socialist country, there was no need to
point out the potential difference between socialism as it developed in the

USSR and socialism as it might develop elsewhere. But as numerous participants pointed out, the existence of the people's democracies of Eastern Europe
and the People's Republic of China posed significant challenges to developing
a universal socialist theory. The authors of the textbook had simply added two
chapters at the end of the textbook to explain the revolutions and socialist
states of Eastern Europe and Asia. Discussion participants wanted to know
why the sections on China and Eastern Europe had not been integrated into
the textbook's section on the transition to socialism. Others asked if each of
the people's democracies of Eastern Europe needed to emphasize heavy industry as was done in the USSR. And how would the USSR's support of Eastern
Europe affect the development of socialism there? Similarly, how did the inter-
dependence of the socialist camp alter socialism's development?

A general consensus quickly formed that the book was "Eurocentric" and
examples from other areas of the world could be used to illuminate each section of the book. One participant pointed out that the textbook would be
read all over the world, and thus diverse examples would help readers understand the book's relevance. Many participants demanded a more thorough
explanation of the Chinese Revolution. Was it a bourgeois-democratic or
some sort of "mixed class" revolution? Why did the book ignore the "crisis of
colonialism" and the process of decolonization?[67]

While many participants criticized, few offered solutions. Yet again the authors of the textbook were put in a position of answering fundamental theoretical questions before they had been debated and resolved by specialists. And
again, the specialists were unlikely to address such questions in print, because
by implication any answer they came up with would have broad political consequences. Was Soviet socialism a rigid model for building socialism elsewhere, or did circumstances in Russia's past make its path to socialism unique?

Questions of modern capitalism were also hotly disputed.[68] Two issues in
particular, the deepening general crisis of capitalism and the inevitability of
war between imperialist countries, encouraged debate. Some accused the authors of the textbook of ignoring the reactionary role of the capitalist political
superstructure during the period of imperialism and the general crisis of capitalism."[69] A. I. Kats went so far as to argue that Lenin's thesis that capitalism
would continue to grow during the period of imperialism clearly contradicted
facts that showed the slowing down of the economy that "is accompanying
the current bourgeois crisis."[70] According to Kats, Lenin's ideas about the general crisis of capitalism were relevant only for the earlier period of imperialism.
Other speakers dismissed Kats's ideas while offering their own critiques of the
textbook's description of the crisis of capitalism. Varga and M. I. Rubinshtein,
for instance, pointed out the need for integrating the postwar militarization
of capitalism into existing theory.[71] While participants generally agreed with
Varga on that point, they could not accept his "revisionist" understanding
about the origins of the general crisis of capitalism. Orthodoxy held that the

crisis began with the October Revolution in Russia, when the advent of social-ist economies limited capitalist expansion and brought imperialist powers in-creasingly into direct competition with one another for goods and markets. Varga, in contrast, asserted that the crisis began sometime before the First World War, when imperialism reached its apex. On this issue Varga stood alone, but since he was the USSR's foremost expert on capitalism, it was clear that the authors of the textbook had to take even his maverick positions seriously.[72]

Varga caused even more of a stir by suggesting that the Party needed to help economists answer the question: "Is Lenin's thesis on the inevitability of war between imperialist countries valid today?" He argued that the thesis had become outdated for three reasons. First, the experience of the two world wars had taught imperialist powers that war between them hurt their eco-nomic strength. Second, imperialist countries were increasingly becoming part of a single military union with common interests that outweighed any internal contradictions in their camp. Finally, the United States had become so dominant that it was difficult to imagine why it would need to go to war with other capitalist powers in order for it to achieve its goals. Others, how-ever, adamantly disagreed.[73] Varga was eager to push political economy—with the Party's permission—beyond the rigid adherence to existing Marxist-Leninist doctrine.

While disagreements over substantive questions of the political economy of socialism and capitalism dominated the discussion, a number of debates about the overall tone of the textbook also arose. Just as Aleksandrov's *History of Western European Philosophy* was criticized for lacking a "militant, Party" spirit, so too did the political economy textbook come under attack for its "scholarly" and "objective" tone. To be fair, the vast majority of participants had no idea that Stalin had implored the authors to "toss out" the propaganda and avoid "abusive language" because political economy was "serious work." In fact, it was not uncommon for speakers to call for a more engaged tone and to explicitly cite Zhdanov's speech at the philosophy discussion to justify their distaste for the distant, academic style of the textbook. The book was supposed to be "a powerful ideological weapon for educating the masses and for battling idealist, bourgeois ideology."[74] M. I. Smit, a researcher at the Institute of Eco-nomics, was confident that the Central Committee wanted students to learn more than the basic laws of the contradictions of capitalism. They must also "learn to hate capitalism." Given this standard, Smit and others declared the textbook "apolitical," "too calm," "pale," "dry," and lacking in "Bolshevik pas-sion." In some of the participants' minds, there was no contradiction between the call for greater scientific standards and the need for more political engage-ment from the authors. As Trakhtenberg put it, the textbook needed "to ap-peal both to the cold logic of intelligence and the hot passion of the heart."[75] One speaker, displaying a bit of passion himself, gave the following recommendation: "I do not think the textbook should be feminine; it should be a masculine, powerful weapon of the democratic anti-imperialist camp." The stenographic record reports laughter in the hall.[76]

Perhaps no one was in a better position to judge the tone of the textbook than the director of the Institute of Philosophy, Aleksandrov, whose own book had been the subject of great ridicule four years earlier. Aleksandrov called for the integration of the ideas of Soviet historians and philosophers as well as Stalin's statements on linguistics. This would aide the political economy textbook in reaching the high standards of "Marxist science." While political economy was based on "objective facts," it also needed to actively engage "the dogma of contemporary bourgeois economic theories." Echoing what had been said about his own book in 1947, Aleksandrov called for the authors of the textbook to seek out methodological battles with their Western counterparts; the textbook could not shy away from its role as a weapon on the economic front. Somehow the authors were supposed to strike the correct balance, pursuing "scientific" explication of "objective truths" while avoiding the trap of apolitical and disengaged scholarship.[77] The fact that Aleksandrov had failed to do just that in his own scholarship did not keep him from demanding it of others.

Participants often echoed Stalin's articles on linguistics and blamed an "Arakcheev regime" in economics for problems with the textbook. Many voiced disapproval of the lack of open discussion among economists and believed that the Institute of Economics was at fault. (One participant even humorously asked if the emphasis on "free and open" discussions implied that there were other kinds taking place among Soviet scholars.) As one speaker noted, "If we have not sufficiently worked out and clarified difficult questions, is it really possible to write a definitive textbook on political economy?"[78] Others lamented poor leadership in the field, the lack of self-criticism, and the endless meetings without any productive results. The Institute of Economics was a "scientific traffic jam" overwhelmed by an atmosphere reminiscent of Oblomov, the do-nothing character from Goncharov's nineteenth-century novel. Some suggested dividing the institute into four or five smaller institutes in the hope of increasing productivity.[79]

The reasons for stagnation in the field did not simply result from a monopoly at the top of the field or a lack of open discussion among economists. All agreed that fear of making a mistake hindered scholarly exchanges and dissuaded people from publishing their work. One participant noted, "The situation has evolved to the point where we wait for Stalin to decide our problems instead of using Stalin's work as a basis for . . . moving our science forward ourselves." If Stalin would soon pass judgment, scholars were reluctant to go out on a limb. Conservatism in the field was compounded by bureaucratic mechanisms that slowed down publication. As one speaker noted, in order to get something published it had to pass through seven or eight levels of review.

Approval was needed from the department, special editors, the institute's administration, the institute's Scientific Council, the editorial committee of the section of the Academy of Sciences, and a special committee of the Academy of Sciences publishing house. Each step of the way, the work could be changed, edited, or rejected. As a last step, the appropriate ministry would pass judgment on whether the work could finally get published.[80] So much for Stalin's calls for "free and open" exchange of scholarly opinions. Given the backlog of unpublished works, the authors of the textbook could not be blamed for not being up to date on the latest ideas in the field.

While most of the participants outlined major theoretical shortcomings in the textbook, some pointed to practical problems in the figures and data available to the authors. Insufficient data hampered both the sections on modern capitalism and on socialism. "Facts, especially data, are the lifeline of science," a specialist in measuring productive power declared. Yet access to the appropriate ministries was denied to many economists, he argued, even though similar access was taken as a given for physicists, chemists, geologists, and other scientists. Iudin also sought to pass some responsibility for the weakness of the textbook on "practical questions" on to the ministries: "Though we consulted with . . . Gosplan and the Ministry of Finance on some occasions, we culled the important information on the Soviet economy from literary sources." Inevitably, these sources were also written without access to Gosplan or the Ministry of Finances statistics.[81] The question of access to reliable data was another way the discussion of the textbook revealed deeper problems in the administration of economic research and publication.

Original plans had called for the discussion to last six or seven days, but Malenkov's opening comments on November 10 suggested it would last twice that long. No one seemed prepared for the meeting to last over five weeks. The discussion ended up including twenty-one plenary sessions with over 110 speeches. By late November, when the meeting was supposed to have ended, some participants appeared almost embarrassed by the repetitiveness of their talks. As one scholar put it, it was becoming difficult to speak and no less difficult to listen. Others were more enthusiastic, remarking that although they had been meeting for over two weeks, "not all questions" concerning the textbook had been raised or addressed. Even Suslov, who began chairing some of the meetings for Malenkov midway through the discussion, lost his patience with long-winded speakers. Recognizing the problem, one speaker got laughs for opening his address, "The work of our conference somehow seems to be slowly approaching completion."[82] It is not exactly clear why the meeting did not simply end. One possibility is that the participants were waiting for Stalin to conclude the discussion with a speech of his own.[83] Though this theory has been neither corroborated nor refuted by archival evidence, it does fit the circumstances. Malenkov, unlike Andrei Zhdanov and Stalin, was not likely to enter into intellectual disputes. And since the authors themselves had consulted with Stalin about the textbook, it followed that he would be the logical person to bring the discussion to a definitive conclusion. But Stalin chose not to give a speech to the meeting.

By early December Malenkov must have realized that things had to start winding down, and he finally gave the authors the floor. Ostrovitianov spoke first, on December 3. He accepted the criticism of the textbook, declaring that the discussion marked a decisive moment in the development of economics in the USSR. First, he recognized that the textbook had not been entirely clear on the economic laws of socialism. On the one hand, laws could be portrayed as natural and uncontrollable, and thus people were subordinated to their powers. This led to mechanistic understandings of political economy, where laws simply dictated human development. On the other hand, if laws were seen as voluntary, with the state and Party capable of creating and destroying them, this would result in anti-Stalinist positions. After all, Stalin had reiterated in his articles on linguistics that the political superstructure cannot control the economic base. Ostovitianov put the problem succinctly: if the state's will is supreme in the economy, "what's left [for economists] to study?" But shying away from the tough questions was not what Ostrovitianov recommended. Admitting that at times the law of value is like "an open electrical outlet that shocks anyone who is brave enough to touch it," Ostrovitianov believed the discussion had helped the authors come to terms with the problems in this area.[84]

As the director of the Institute of Economics, Ostrovitianov admitted that the harsh criticism of the situation on the "economic front" was justified and that the institute had not done enough to organize discussions and publish work. Malenkov interrupted him to ask, "Why aren't books getting published? What's going on?" Ostrovitianov responded with a general statement to the effect that he had not taken to heart Stalin's words that "no science can develop without the battle of opinions and free criticism." Unsatisfied, Malenkov pushed him on the question. Ostrovitianov noted that there was a big difference between the nonpublished disagreements economists had about theoretical questions and those that found their way into print. If something was subject to debate, it was not printed. The primary concern of economists had been to avoid making a mistake in print. This, Ostrovitianov now understood, had been wrong. He also blamed the cumbersome review process for delaying publications and editing work to such an extent that authors might not even recognize their own articles. He suggested that he should have approached the Central Committee about this problem. Finally, he added that economic literature suffered from lack of access to sources and data contained at Gosplan and in economic ministries.[85]

Three days later, Leont'ev and Pashkov addressed the meeting. Leont'ev spent most of his talk defending the "historical method." Some participants had suggested that the textbook should follow the "analytical method" used by Marx in *Capital*. By beginning the section on socialism with definitions, the confusion over the various stages of socialism might be avoided. Leont'ev rebutted this argument by saying that the historical approach allowed him to avoid scholasticism and abstractions. Though he did not mention it in his talk, Stalin had told him to use a historical approach, and he was not about to abandon it, no matter what its weaknesses. Leont'ev admitted that there were errors in the textbook and thanked the comments of specialists who pointed out problems with specific aspects of the book. He fully accepted the need to integrate modern colonialism into the section on monopoly capitalism and agreed that the Eurocentric approach diminished the ideological value of the book. There was some criticism he could not agree with, however. He took issue with Varga's idea that the general crisis of capitalism predated the October Revolution. He also did not agree with those that thought the book should be written in a more "militant" tone, emphasizing, "We will not integrate the shrieking and vulgar style of cheap, political agitation into the textbook." Though he admitted that there were some sections where they had not taken the opportunity to assist the "working class in digging capitalism's grave," he insisted that "political economy is not civic history," where politics and propaganda were more appropriate. He ended his talk, much the way Ostrovitianov began his, with the declaration that the discussion marked a complete revolution for economics in the USSR.[86]

Iudin, the last author to address the meeting, was the most divisive. He accused Ostrovitianov's institute of not following the Central Committee's lead in holding discussions where everyone was allowed to speak their mind. He also agreed that the role of the Soviet state had not been sufficiently emphasized in the textbook. Iudin accepted that the state could not determine all economic laws. After all, if things were that simple, the USSR could have solved all its economic problems in two or three years. Just the same, Iudin emphasized that the USSR played a bigger role in its economy than any other state in history. Iudin identified a central paradox. Those who praised the Soviet state for its all-powerful role in the economy must also hold the state responsible for any shortcomings in the economy. For a dialectician like Iudin the solution to the paradox was not difficult. The state and Party were indeed quite powerful, but their power came from the will of the people rather than any economic laws. Just because Party politics played a role in industrialization, collectivization, and the building of socialism, it did not follow that Party politics could be equated with economic laws. Instead, economic laws set parameters on state power and could thus be blamed for any instances where the Soviet economy seemed to struggle. Wisely, Iudin did not give any examples of Party politics and economic laws contradicting one another, or any indication of what should be done in those situations.[87]

As the last person to join the group of authors, Iudin distanced himself from the textbook. He agreed with those who criticized the closed and elite atmosphere surrounding the production of the textbook: "In the course of the discussion many comrades persistently asked the question why over a 15-year period economists have not been able to write a political economy textbook. The question is completely legitimate. I think that this question would be best answered by comrade Leont'ev. (commotion in the hall) All of the participants here know that for 15 years comrade Leont'ev has in fact monopolized the preparation of the political economy textbook. I wanted to point out this problem with comrade Leont'ev's [lack of] self-criticism."[88] Leont'ev was not given the opportunity to defend himself, and despite clear plans to continue the plenary sessions, Iudin's divisive note proved to be the last word. Malenkov reported to Stalin that 116 economists had spoken and that dozens more were waiting to address the meeting.[89] In the hope of establishing some concrete suggestions, he decided to break the meeting into simultaneous sessions. After December 8, the economists met for nine days in smaller groups to discuss capitalism and socialism. These meetings proved less formal, but for the most part participants simply raised questions that had already come up during the plenary sessions. They met for the last time on December 17 and, without any fanfare or even the simplest of closing remarks, the 250 participants were finally sent home.[90]

Even before the discussion sputtered to a close, economists and Central Committee bureaucrats began to worry about presenting the results to Stalin in the best possible light. On December 8, 1951, D"iachenko sent a long report to Malenkov defending himself personally and the Institute of Economics more generally from attacks that had been leveled during the course of the discussion. Similarly, Leont'ev penned a detailed letter to Malenkov rebutting Iudin's suggestions that he had monopolized work on the textbook for fifteen years.[91] Sometime before the last plenary meeting Malenkov brought together a small group of discussion participants under Suslov and Yuri Zhdanov to draft a proposal for improving the textbook. By December 22 the group had pulled together three detailed reports based on material from the discussion. The first report, "Proposals for Improving the Textbook on Political Economy," described the substantive inadequacies of the textbook. The second report, "Proposals for the Elimination of Factual and Editorial Mistakes and Inaccuracies," compiled sentence-by-sentence corrections to the original text as they had been proposed by the participants in the discussion. Finally, "Index to Controversial Questions Arising during the Discussion of the Textbook on Political Economy" outlined unresolved debates concerning the content of the textbook. A fourth report, compiled by the Central Statistical Administration, criticized the use of data in the book and provided updated information. Each of the four reports was substantial enough to be bound as a small booklet, and each offered a level of detail that far surpassed Agitprop's previous excursions into scholarly debate.[92]

On December 22 Malenkov, Suslov, and Zhdanov drafted a report to Stalin along with a cover letter outlining the organizational structure, tone, and content of the monthlong discussion. They suggested that the authors present a completed copy of the revised textbook to the Central Committee by May 1, 1952. Four economists, L. M. Gatovskii (acting chairman of the section on the political economy of socialism of the Institute of Economics), I. I. Kuz'minov (an instructor at the Central Committee's Academy of Social Sciences), V. I. Pereslegin (chief accountant of the Ministry of Finance), and A. M. Rumiantsev (a researcher at the Institute of Economics who had previously worked in the Science Section of Agitprop), were chosen to supplement the existing group of authors. Accompanying the report, a draft resolution of the Central Committee outlined the problems with the textbook and ordered the authors to use the materials from the discussion to improve the textbook for resubmittal to the Central Committee.

Malenkov, Suslov, and Zhdanov also reported to Stalin on the generally poor situation in economics that had become evident in the course of the discussion. Numerous theoretical problems concerning the economics of capitalism and socialism were compounded by the lack of "criticism" and "open discussions" among economists. The report informed Stalin that "in recent years, the Institute of Economics . . . has not completed a single serious scientific project . . . and has not informed the Council of Ministers and the Central Committee of developments in Soviet and foreign economies." The report held Ostrovitianov personally responsible for the poor state of affairs at the institute and blamed systemic "excessive caution" for the backlog of unpublished manuscripts. Among other things, a draft Politburo resolution called for Ostrovitianov's dismissal and replacement by the former Central Committee apparatchik Rumiantsev.[93]

Participants in the discussion also received copies of the official reports about the meeting's findings. Many of them were not pleased with the way Suslov, Zhdanov, and others had summarized their views, and for the next few months they wrote to Malenkov clarifying their positions. Since their letters were simply forwarded to Zhdanov, they were not likely to get anywhere with their complaints. Zhdanov's assistants at the Central Committee even filed Leont'ev's suggestions for improving the reports without taking any action.[94]

Meanwhile, Stalin's intentions were unknown. Would he publish the minutes of the discussion according to the original Central Committee plan? This seemed unlikely because the discussion had lasted much too long and ended unceremoniously. The minutes, which revealed more contentiousness than consensus, would have filled at least three large volumes. But declaring the discussion a failure or simply ignoring it was also risky. After all, beginning the textbook once again from scratch would have added years to the project.

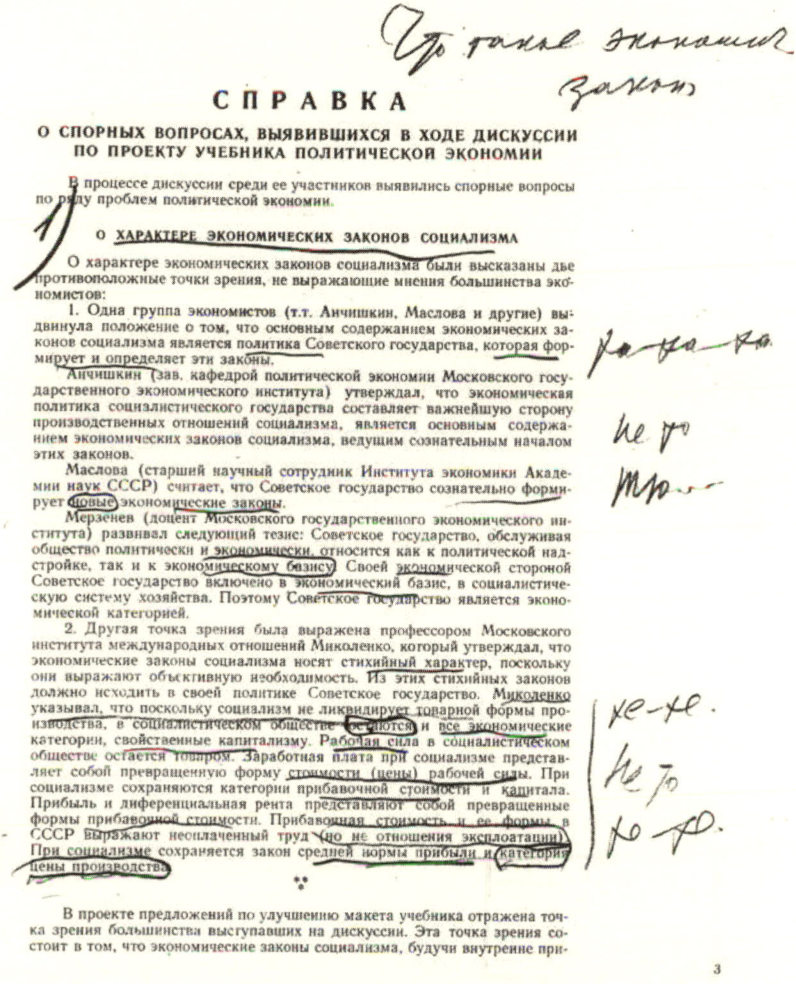

[Handwritten marginalia at top:] Про такие экономич. законы

СПРАВКА

О СПОРНЫХ ВОПРОСАХ, ВЫЯВИВШИХСЯ В ХОДЕ ДИСКУССИИ ПО ПРОЕКТУ УЧЕБНИКА ПОЛИТИЧЕСКОЙ ЭКОНОМИИ

В процессе дискуссии среди ее участников выявились спорные вопросы по ряду проблем политической экономии.

О ХАРАКТЕРЕ ЭКОНОМИЧЕСКИХ ЗАКОНОВ СОЦИАЛИЗМА

О характере экономических законов социализма были высказаны две противоположные точки зрения, не выражающие мнения большинства экономистов:

1. Одна группа экономистов (т.т. Анчишкин, Маслова и другие) выдвинула положение о том, что основным содержанием экономических законов социализма является политика Советского государства, которая формирует и определяет эти законы.

Анчишкин (зав. кафедрой политической экономии Московского государственного экономического института) утверждал, что экономическая политика социалистического государства составляет важнейшую сторону производственных отношений социализма, является основным содержанием экономических законов социализма, ведущим сознательным началом этих законов.

Маслова (старший научный сотрудник Института экономики Академии наук СССР) считает, что Советское государство сознательно формирует новые экономические законы.

Мерзенев (доцент Московского государственного экономического института) развивал следующий тезис: Советское государство, обслуживая общество политически и экономически, относится как к политической надстройке, так и к экономическому базису. Своей экономической стороной Советское государство включено в экономический базис, в социалистическую систему хозяйства. Поэтому Советское государство является экономической категорией.

2. Другая точка зрения была выражена профессором Московского института международных отношений Миколенко, который утверждал, что экономические законы социализма носят стихийный характер, поскольку они выражают объективную необходимость. Из этих стихийных законов должно исходить в своей политике Советское государство. Миколенко указывал, что поскольку социализм не ликвидирует товарной формы производства, в социалистическом обществе сохраняются и все экономические категории, свойственные капитализму. Рабочая сила в социалистическом обществе остается товаром. Заработная плата при социализме представляет собой превращенную форму стоимости (цены) рабочей силы. При социализме сохраняются категории прибавочной стоимости и капитала. Прибыль и диференциальная рента представляют собой превращенные формы прибавочной стоимости. Прибавочная стоимость и ее формы в СССР выражают неоплаченный труд (но не отношения эксплоатации). При социализме сохраняется закон средней нормы прибыли и категория цены производства.

В проекте предложений по улучшению макета учебника отражена точка зрения большинства выступавших на дискуссии. Эта точка зрения состоит в том, что экономические законы социализма, будучи внутренне при-

3

Figure 14. Stalin's marginalia, written in early 1952, on the "Index to Controversial Questions Arising during the Discussion of the Textbook on Political Economy." Stalin wrote "Ha-Ha-Ha" and "Not so" at points where the report summarized views expressed by participants in the discussion. Courtesy of RGASPI.

Given the goal of producing a definitive textbook quickly, Stalin chose once again to get directly involved.

Soon after receiving the records of the meeting, Stalin began to pore over them. He read through the speeches quickly, marking up some of them heavily and skipping others altogether.[95] He seems to have read with a pencil in his hand. When he studied the index to controversial questions that had arisen

сущими социалистическому способу производства, представляют собой неизбежный результат развития материальной жизни общества, а не произвольный продукт сознания и воли людей. Государство, партия не могут создавать, формировать экономические законы. Политика партии и государства исходит из законов развития производства, из экономических законов развития общества.

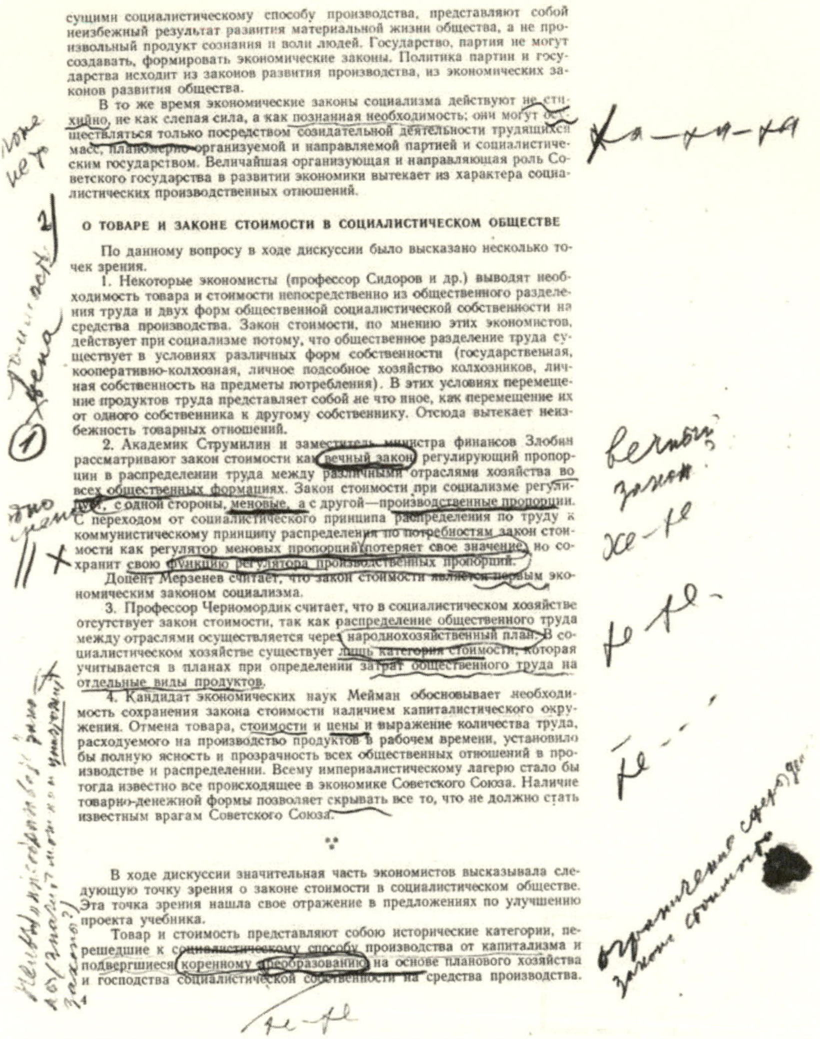

В то же время экономические законы социализма действуют не стихийно, не как слепая сила, а как познанная необходимость; они могут осуществляться только посредством созидательной деятельности трудящихся масс, планомерно организуемой и направляемой партией и социалистическим государством. Величайшая организующая и направляющая роль Советского государства в развитии экономики вытекает из характера социалистических производственных отношений.

О ТОВАРЕ И ЗАКОНЕ СТОИМОСТИ В СОЦИАЛИСТИЧЕСКОМ ОБЩЕСТВЕ

По данному вопросу в ходе дискуссии было высказано несколько точек зрения.

1. Некоторые экономисты (профессор Сидоров и др.) выводят необходимость товара и стоимости непосредственно из общественного разделения труда и двух форм общественной социалистической собственности на средства производства. Закон стоимости, по мнению этих экономистов, действует при социализме потому, что общественное разделение труда существует в условиях различных форм собственности (государственная, кооперативно-колхозная, личное подсобное хозяйство колхозников, личная собственность на предметы потребления). В этих условиях перемещение продуктов труда представляет собой не что иное, как перемещение их от одного собственника к другому собственнику. Отсюда вытекает неизбежность товарных отношений.

2. Академик Струмилин и заместитель министра финансов Злобин рассматривают закон стоимости как вечный закон регулирующий пропорции в распределении труда между различными отраслями хозяйства во всех общественных формациях. Закон стоимости при социализме регулирует, с одной стороны, меновые, а с другой—производственные пропорции. С переходом от социалистического принципа распределения по труду к коммунистическому принципу распределения по потребностям закон стоимости как регулятор меновых пропорций потеряет свое значение, но сохранит свою функцию регулятора производственных пропорций.

Доцент Мерзенев считает, что закон стоимости является первым экономическим законом социализма.

3. Профессор Черномордик считает, что в социалистическом хозяйстве отсутствует закон стоимости, так как распределение общественного труда между отраслями осуществляется через народнохозяйственный план. В социалистическом хозяйстве существует лишь категория стоимости, которая учитывается в планах при определении затрат общественного труда на отдельные виды продуктов.

4. Кандидат экономических наук Мейман обосновывает необходимость сохранения закона стоимости наличием капиталистического окружения. Отмена товара, стоимости и цены и выражение количества труда, расходуемого на производство продуктов в рабочем времени, установили бы полную ясность и прозрачность всех общественных отношений в производстве и распределении. Всему империалистическому лагерю стало бы тогда известно все происходящее в экономике Советского Союза. Наличие товарно-денежной формы позволяет скрывать все то, что не должно стать известным врагам Советского Союза.

В ходе дискуссии значительная часть экономистов высказывала следующую точку зрения о законе стоимости в социалистическом обществе. Эта точка зрения нашла свое отражение в предложениях по улучшению проекта учебника.

Товар и стоимость представляют собою исторические категории, перешедшие к социалистическому способу производства от капитализма и подвергшиеся коренному преобразованию на основе планового хозяйства и господства социалистической собственности на средства производства.

Figure 15. Here Stalin continued with his derisive remarks. After taking his pen to the various reports from the discussion, Stalin began to write an essay of his own. Courtesy of RGASPI.

during the discussion he evidently became more agitated. In the margins he wrote "that's not true," "nonsense," "ha-ha-ha," "what is this?" "stupidity," and so on. He also began to write longer ideas: "Political economy is not the history of the revolution," and "These are all political themes, but we need economic reasons for the failure of capitalism."[96] At some point in early 1952 he decided to write up his comments separately.

Stalin's essay, titled "Remarks on Economic Questions Connected with the November 1951 Discussion," was addressed to the participants in the discussion. Logically, Stalin referred to the discussion throughout the text, mentioning the opinions of the "majority of participants," remarking on the degree to which the textbook was criticized at the discussion, and passing judgment on the necessary steps to complete the textbook. Though Stalin's essay was subsequently published, the discussion to which Stalin was referring remained obscure to all but the organizers and participants. This has led to some confusion about what Stalin was trying to do. Archival access to Stalin's papers now clarifies that the essay's primary purpose was to address controversial issues that had been raised at the November 1951 meeting of economists.

Stalin began the "Remarks" with a statement about the character of economic laws under socialism. He emphasized that economic laws in all economic formations were "objective" and "scientific." By this he meant that economic laws, like the laws of the natural sciences, could not be created, destroyed, or transformed by human will. Some participants in the discussion had emphasized the role of the Soviet state in dictating economic laws. Stalin disagreed. He admitted that the Soviet state had the difficult tasks of replacing the exploitative imperial Russian economy and creating the basis of a socialist economy through rapid industrialization, but he insisted that it was the utilization of economic laws that brought about these changes in Soviet society. For instance, the "law of the planned proportional development" of the economy replaced the competition and anarchy of the market indicative of capitalism. This law made it possible for the Soviet state to determine which sectors of the economy needed support or emphasis. Planning reflected economic laws, not political will. Stalin declared that those who denied that the "processes of economic life are law-governed and operate independently of our will" were in fact "denying science . . . and the possibility of directing economic activity."[97]

Given that economic laws existed (as had been the consensus since Stalin's 1941 conversation with economists), participants in the discussion had wanted to know what were the basic economic laws of modern capitalism and socialism. Some had suggested that the basic economic law of capitalism was the law of value, while others had suggested that it was the law of competition and the anarchy of production. Stalin felt obliged to share his views on the basic economic laws of capitalism, which he summed up as follows: "the securing of the maximum capitalist profit through the exploitation, ruin and impoverishment of the majority of the population of the given country, through the enslavement and systematic robbery of the peoples of other countries, especially backward countries, and lastly, through wars and militarization of the national economy, which are utilized for the obtaining of the highest profits." The basic law of socialism, in contrast, was formulated as "the securing of the maximum satisfaction of the constantly rising material and cultural requirements of the whole of society through the continuous expansion and perfection of socialist production on the basis of higher techniques."[98]

One might assume from these definitions that the economic laws governing socialism would have nothing in common with the economic laws governing capitalism. For instance, some participants had argued that commodity production should be eliminated in the USSR. Stalin agreed that if all the means of production had been seized by the state on behalf of the proletariat, then commodity production would be unnecessary in the USSR. But so long as the collective farms were not publicly owned and their products were exchanged with state-owned industry, money and commodity circulation would persist. This did not mean that the state could socialize the collective farms' means of production: "Marxists cannot adopt this senseless and criminal course, because it would destroy all chances of victory for the socialist revolution, and would throw the peasantry into the camp of the enemies of the proletariat for a long time." In other words, no massive expropriation of collective farm property was advisable or imminent.[99]

Did the persistence of commodities in socialism pose a danger of the re-emergence of capitalism? Stalin argued that it did not, because the public ownership of the means of production prevented the unrestrained spread of commodity production. Still, it followed that if there were commodities in the Soviet Union, then there also had to be the law of value. According to Stalin, the law of value served two functions: first, in a limited sense, it regulated commodity circulation by assigning values to products exchanged between the various sectors. Second, it influenced production, without actually regulating production. The law of value assisted individual factory managers in increasing their efficiency and in calculating levels of profits. But because the plan regulated the overall economy, overproduction was avoided, and even the less profitable industries were kept afloat. Thus the laws of value and profit were subordinated to the concept of the planned proportional development of the economy.[100]

Stalin's "Remarks" reveal a certain willingness on the leader's part to reconsider earlier formulations or interpretations of his work. Following Marx's lead, Stalin had talked about the abolition of the distinctions between town and country and between mental and physical labor. He now emphasized that while the *essential* distinctions would disappear, some distinctions would naturally remain. Admitting that his earlier formulation was "imprecise and unsatisfactory," Stalin suggested that it should be discarded and replaced by a new statement acknowledging the persistence of inessential distinctions.[101] He was also willing to revise official doctrine on the deepening crisis of capitalism. He admitted that his prewar thesis stating that markets would remain relatively stable during the general crisis of capitalism could no longer be considered valid. Instead, the loss of such markets as China and the "disintegration of the world market" were contributing to the deepening of the crisis of capitalism. Likewise, the new pressure on the capitalist economy meant that Lenin's 1916 thesis that "capitalism was growing more rapidly" had also "lost its validity."[102] There were, of course, limits to such revisions. Stalin supported Lenin's assertion of the inevitability of wars between capitalist countries: Varga was wrong. So long as there was capitalism and imperialism, the struggle for markets would outweigh any potential cooperation between capitalist countries.[103]

Stalin ended his essay by emphasizing the importance of the textbook for communists in the USSR and around the world. He warned against the temptation to include too much in the textbook and outlined his ideas for bringing about the successful completion of the book. He also suggested that even some "out and out critics" of the draft be included on a small committee to oversee the rewriting and editing of the book. This committee was given one year to present the finished product to the Central Committee.[104]

On February 7, 1952, five days after completing his "Remarks," Stalin spoke with Ostrovitianov on the telephone. According to Ostrovitianov's hastily written notes, Stalin called him at 1:00 a.m. to inform him that he had written "about fifty pages" in response to the discussion of the textbook. The two of them concluded that ten to fifteen people should be brought together to discuss the essay and agreed that it was "probably not expedient to gather all the participants in the discussion." When Ostrovitianov asked Stalin if his "Remarks" could be published, Stalin responded, "No. This is not for publication. Publishing them would not be in your interest. The remarks were not ratified by the Central Committee in order not to hamper the work of the authors of the textbook." Within a few minutes, Malenkov called with follow-up questions for Ostrovitianov, and they agreed that all the participants in the discussion should receive a copy of Stalin's "Remarks."[105] In early February, Stalin's essay was distributed to all the participants in the discussion and all members of the Central Committee.

By February 12 the Central Committee was flooded with requests from scientific institutes, educational establishments, regional and republican apparatchiks, and others asking for copies of Stalin's "Remarks." Central Committee members and the discussion participants must have told other people that Stalin had reviewed the discussion materials. Yuri Zhdanov and his deputy in the Science Section wrote to Malenkov recommending that the Central Committee distribute copies to political economy teachers, political economy departments, academy institutes, the editors of major journals and newspapers, and secretaries and scientific workers of the republican, regional, and city Party cells. The total number of additional copies came to about three thousand. They were supposed to be sent with instructions explaining that "citing the 'Remarks' in the press is not allowed until it has been published."[106] It is uncertain whether Malenkov followed Zhdanov's recommendation, but it is clear that knowledge of Stalin's important "theoretical state-ment" was spreading quickly.

On February 15, 1952, a select group of seventeen economists met with Stalin and members of the Politburo to hear firsthand Stalin's ideas about political economy.[107] The conversation that subsequently took place between Stalin and the economists reveals the extent to which the scholars had be-come dependent on the political leader to address problems in their field. They asked him even about minute details, such as whether a term should be hyphenated or whether a quotation mark should surround a phrase. If Stalin introduced an unfamiliar phrase in his "Remarks" or suggested a new line of inquiry, the economists wanted to make absolutely sure that they understood its implications entirely. Perhaps this is why Ostrovitianov opened the conver-sation by essentially repeating his question about whether the "Remarks" should be published and how scholars should use the essay in their teaching and scholarship. Stalin again insisted that the "Remarks" should not be pub-lished, adding, "The political economy discussion was closed and the people don't know about it. The speeches of the participants in the discussion were not published. People will not understand if I appear in the press with my 'Remarks.' Publication of the 'Remarks' in the press is not in your interest. They will understand that everything in the textbook was determined in ad-vance by Stalin. I'm worried about the authority of the textbook. The text-book should have undisputed authority. It would be right if the things in the 'Remarks' were first known from the textbook."[108] Stalin suggested that the economists use the "Remarks" in lectures and "political circles" without quot-ing the author. It would only be appropriate to publish his essay a number of years after the publication of the textbook. Showing remarkable insight into the paradox he had created, Stalin sensed that his official proclamations, if widely publicized, might diminish the authority of the textbook.

Ostrovitianov also asked for clarification of the term "consumer commodi-ties" and whether it could be applied to the means of production. As he under-stood it, "consumer goods" was a necessary category for those managing part of the economy producing the means of production. Stalin countered that the means of production were not freely bought and sold in the USSR and there-fore could not be considered similar to commodities. Did this mean that the means of production were a "commodity of a special type?" Stalin said no. Managers working in those spheres used the law of value to make calculations and for "checking the advisability of an action," but since the means of pro-duction were themselves not freely bought and sold, they could not be consid-ered a commodity at all.[109] Ostovitianov's final question concerned Stalin's use of the phrase "the crisis of the world capitalist economic system." Was this somehow different from the "general crisis of capitalism"? Stalin answered that the two phrases were interchangeable but that he believed it was necessary to emphasize the capitalist system as a whole, organic unit and to dissuade scholars from examining only the economies of individual capitalist countries.[110]

On the question of the role of economic laws in socialism, Stalin repeated that laws could not be created or destroyed and therefore the notion of a "transformed" law made no sense. He added, "If you can transform and abolish a law of science, this means that we are good for nothing. Laws must be considered, controlled, and used. The sphere in which they apply can be limited. This is so in physics and chemistry. This is so in relation to all science."[111] A logical follow-up question was asked a few minutes later. If there were objective laws in economics, how did the Soviet state interact with those laws? Was the state part of the base, meaning that it was an integral part of economic laws, or was it part of the political superstructure acting in response to the economic base? Stalin understood that the question directly referred to his articles on linguistics and laughed, remarking that a lot had been said on the topic and that some people "even understand Soviet power as part of the base." Inasmuch as the Soviet budget was distinct from capitalist state budgets, it was part of the base. But there were also "elements of the superstructure" in state finance, such as decisions about expenditures. Because communism had not yet been reached, the state still made political decisions based on the needs of the whole economy. The state thus used the economic law of "planned proportional development" to coordinate various sectors of the economy. This essentially meant that economists had to be "objective scientists" studying economic "laws," but they also could not turn a deaf ear to political concerns. The field was still theoretically a hybrid between science and politics.[112]

The superstructural role of the state would persist until full communism had been achieved. For those who believed that communism was around the corner, Stalin had sobering news. He reminded the economists that "the transition to communism requires solutions to a mass of questions." He believed that the "peasants themselves" had to want to leave the agricultural artels, but it was clear that they were not yet concerned enough with broad "societal affairs" to make the transition. He explained that so far the "kolkhoz worker doesn't think about anything but himself and doesn't want to know anything of economics" and added that "it is impossible to imagine the transition to the second phase of communism through narrow-mindedness. No 'special' step to communism will happen. Slowly, without our noticing, we will enter communism." The road to communism was long, and the leader emphasized that "it is not advisable to rush . . . there is no need to hurry."[113]

The day after the meeting between Stalin and the seventeen economists, the Politburo passed a resolution creating an eleven-member commission to complete the political economy textbook. In a clear demotion for Leont'ev, Ostrovitianov chaired the commission and became the chief economist responsible for the textbook. The group was given a little over a year to finish the job. The mandated due date of March 1953 was reasonable considering that all the members of the commission were once again freed from all other professional responsibilities.[114]

Meanwhile, the controlled distribution of Stalin's "Remarks" did not settle all the questions in the field. To the contrary, by April 1952 Zhdanov had compiled a list of seventy-two different questions raised by teachers and scholars in response to Stalin's "Remarks." The breadth and depth of the questions suggested that even after Stalin's intervention into the discussion, countless problems remained unanswered. The Secretariat simply passed Zhdanov's list to the authors of the textbook.

Some letters evidently made it to Stalin, however, and the general secretary composed written responses to three of them during the spring and summer of 1952.[115] Stalin first responded to a letter from A. I. Notkin, a senior scholar at the Institute of Economics who had participated in the discussion. Notkin addressed a series of issues that had come up at the discussion, including the ways in which states utilized economic laws and the role of profit in the Soviet economy. Even though he echoed some of the questions that the prominent economists had put to Stalin in February (about the crisis of capitalism and the means of production as commodities in socialism), Stalin was clearly not impressed with the letter. "I was in no hurry to reply, " Stalin wrote, "because I saw no urgency in the questions you raised." Still he addressed Notkin's letter "point by point."[116] Notkin was curious about the utilization and manipulation of economic laws in periods other than socialism. He assumed that, under capitalism, economic laws dictated the actions of state and society and that only under socialism was the state free to use economic laws as it saw fit. Stalin disagreed, arguing that economic laws were used under all economic formations but that in class-based societies the laws were always utilized to benefit the dominant class. In contrast, under socialism the class interests of the proletariat corresponded to the interests of the society as a whole. Notkin may have been pushing for a distinction between economic laws (such as the law of value) that could be manipulated by the state during any period of economic development and those iron laws of Marxism (such as the inevitable victory of socialism over capitalism) that were independent of the will of the state or people. Stalin made no such distinction.[117]

Stalin was more willing to accept distinctions between the types of profit that were possible under socialism. While emphasizing the "higher form" of profitability that came from the socialist planning of the economy, Stalin did not mean to suggest that the profitability of individual plants and factories was not important. The state might choose to support plants that incurred great losses (for example, in order to increase the production of the means of production), but it also must take into account what plants were profitable. Only this would allow the state to plan construction and production successfully.[118] Stalin basically used Notkin's letter to elaborate points he had made in his original response to the discussion. This usually meant dismissing out of hand any apparent contradictions.

Stalin next chose to respond a letter written by L. D. Iaroshenko. In this reply his tone was even more dismissive; he described Iaroshenko's views as "un-Marxian," "profoundly erroneous," "chimerical," and "reminiscent of Bukharin." Such name-calling hardly seems appropriate for a "free and open discussion." In fact, it is fair to conclude that Iaroshenko was being punished precisely for pushing the limits of allowable debate in Stalin's Soviet Union.

Long before Stalin singled out his letter, Iaroshenko had earned a poor reputation among Soviet economists and in the Central Committee. As a Party member who worked at the Moscow Regional Statistical Administration, Iaroshenko displayed a keen interest in questions of political economy even though his colleagues and Central Committee apparatchiks consistently and decisively dismissed his views. Despite three unsuccessful attempts to appeal to Malenkov for assistance in publishing his articles, Iaroshenko wrote to the Central Committee in September 1951 asking to be sent a copy of the draft textbook and to be included in the upcoming discussion. Agitprop recommended that his request be turned down, but Malenkov overruled them and Iaroshenko actively participated in the discussion.[119] In his speech to the plenary session and then again in two speeches in the small section meetings, Iaroshenko challenged the textbook's definition of political economy. First, he declared that economic laws of socialism had nothing in common with the economic laws of capitalism. All talk to the contrary was simply scholasticism and reflected the inability of Soviet economists to accept that the liquidation of private property changed the laws and meaning of political economy. Instead of emphasizing production relations as the textbook did, Iaroshenko believed that Soviet economists needed to rationalize Soviet productive forces and to create a scientific basis for economic planning. He insisted that the rational planning of the whole economy, rather than the study of relations between the different sectors of society, constituted the appropriate goal of Soviet political economists. In socialism, concern with the scientific organization and rationalization of productive forces should have replaced concern with money, commodities, credit, and other capitalist categories.[120]

Other participants in the discussion refuted his views, and the "Index to Controversial Questions" contained no mention of Iaroshenko or his proposals for radically shifting the definition of the political economy of socialism. On January 4, 1952, after receiving a copy of the "Index to Controversial Questions," Iaroshenko wrote to Yuri Zhdanov to complain about being excluded from the report. In a memo to Malenkov and Suslov, Zhdanov insisted that the definition offered by the textbook was not controversial and that Iaroshenko was simply wrong.[121] Zhdanov no doubt understood that Stalin had approved the textbook's definition of political economy during his 1950 meetings with the authors. Even after reading Stalin's "Remarks," however, Iaroshenko believed that his understanding of political economy remained correct. On March 20, 1952, he wrote to the Central Committee once again outlining his views on the definition of the political economy of socialism. This time he was not simply challenging the authors of the textbook; he was disagreeing with Stalin himself. Showing great hubris, Iaroshenko suggested that he could write a better textbook in one year if given the time and a few assistants.[122]

Over the next month the Central Committee reviewed Iaroshenko's previous letters to the Central Committee, his contributions to the discussion, and his political standing. In April Zhdanov sent Malenkov a detailed summary of Iaroshenko's ideas. In May Malenkov received short bios of Iaroshenko from Iaroshenko's boss, from the head of cadres at Gosplan, from a Central Committee inspector, and from his local Party organization. These reports characterized Iaroshenko as stubborn, difficult to get along with, undisciplined, inconsistent, never satisfied, and tactless.[123] Given this background, Stalin must have been particularly anxious to denounce Iaroshenko's views, which he did at the end of May 1952.

Stalin argued that though productive forces in socialism had developed, they were still dependent on the replacement of capitalist production relations in both industry and agriculture. Rather than simply reducing political economy to the rational organization of productive forces, Soviet economists had the much more difficult task of illuminating the relations between various sectors of the Soviet economy. Communism did not mean simply the rationalization of productive forces, as Iaroshenko suggested. Stalin pointed out that Iaroshenko had given no indication, either in his speech at the discussion or in his letters to the Central Committee, of the details such a "rationalization" would entail.

In addressing Iaroshenko's definition of the political economy of socialism, Stalin reached the fever pitch that he had insisted the authors of the textbook should avoid. Stalin labeled Iaroshenko a "retrograde Marxist" spouting "unholy twaddle." In conclusion, Stalin dismissed Iaroshenko's complaints about being ostracized by the discussion organizers and declared Iaroshenko's proposal to write the textbook himself ludicrous. Why Stalin bothered to write a response to Iaroshenko remains a matter for speculation. Perhaps he sensed in Iaroshenko's ideas the remnants—albeit in exaggerated form—of the rational planners of the 1920s who had fallen out of favor with the radical economic break of the late 1920s and 1930s. Now that political economy had reemerged as a legitimate "scientific" discipline, it might have seemed natural to some economists to resurrect debates from the 1920s, replete with disagreements about the relationship between economic experts and political leaders. While Stalin clearly supported the trend toward grounding economics in "scientific laws," he also was careful not to suggest that economists were somehow free from political concerns. In some ways, both Voznesenskii and Varga could be faulted for emphasizing the technical aspects of economics—in the form of statistics about capitalism and socialism—without paying sufficient attention to issues of political import. Within this context, Iaroshenko's letter provided a crude foil for Stalin to denounce the tendency to reduce all economics to technical questions of rationalization and planning.

As would be expected, after Stalin's response Iaroshenko paid a price for his earnest interpretation of "open discussion" and "free criticism." The Moscow Party Committee strongly denounced his "Bogdanovite-Bukharinist" views on questions of political economy, and he was sent to work at an institute in Irkutsk. His troubles evidently followed him there, and in the fall of 1952 Iaroshenko wrote to the Central Committee complaining that he had been separated from his family and loved ones. In December 1952 the Central Committee looked into the matter and determined that Iaroshenko had not reformed in light of the Moscow Party decision and maintained his anti-Marxist views. His wife even wrote to the Central Committee explaining that she had refused to accompany him to Irkutsk and that this explained their separation.[124] After his last letter of complaint, Iaroshenko was once again brought to Moscow, where he met briefly with members of the Secretariat.[125] After the meeting, he was arrested and brought to the Lubianka prison. Nine months after Stalin's death he was released.[126]

Stalin's final statement on political economy came in the form of a response to two more economists on September 28, 1952. In contrast to the tone he had adopted with Iaroshenko, Stalin's reply to A. V. Sanina and V. G. Venzher was much more collegial. He noted that the two authors had made a "profound and serious study of economic problems" and offered "quite a number of correct formulations and interesting arguments." Still he discerned "grave theoretical errors," particularly in their theories about collective farms and the character of economic laws of socialism. Stalin argued that limiting the sphere of commodity circulation was a necessary step to achieving communism. Since the collective farms were producing surplus product that could be sold at market, they were a central reason for the persistence of commodity circulation in the USSR. If that market could be replaced by direct product exchange between industry and the collective farms, it would decrease the sphere of commodity circulation and essentially integrate collective farm property with public property. But Stalin warned that such replacement would require "an immense increase in the goods allocated by the town to the country, and it would therefore have to be introduced without any particular hurry, and only as the products of the town multiply." Again the road to communism was clear, but travel along it would be slow.[127]

Sanina and Venzher's letter also inspired Stalin to address once again the nature of the economic laws of socialism. After repeating that economic laws could not be "created" or "transformed," Stalin explored the implications of what would happen if economic laws under socialism were not "objective."

He concluded that abandoning the idea of the objectivity of economic laws "would lead us into the realm of chaos and chance," adding: "The effect would be that we would destroy political economy as a science, because science cannot exist and develop unless it recognizes the existence of objective laws, and studies them. And by destroying science, we should be forfeiting the possibility of foreseeing the course of developments in the economic life of the country, in other words, we would be forfeiting the possibility of providing even the most elementary economic leadership."[128] By reiterating that scientific laws were objective and beyond human ability to create and destroy, Stalin left little doubt that scholars occupied a realm in Soviet society that existed beyond politics. Even in the field of economics, the most politicized of all Soviet disciplines, the solution to scholarly stagnation was to recognize that scholars served a special function for society. Stalin insisted that socialism be scientific.

Stalin's "Remarks" and his responses to Notkin, Iaroshenko, and Sanina and Venzher were finally published as *Economic Problems of Socialism in the USSR* on October 3, 1952. Two days later the XIX Party Congress opened, and Stalin's new "theoretical masterpiece" dominated the proceedings. The congress passed a resolution calling for a commission, chaired by Stalin, to revise the Party's Program. The textbook authors Rumiantsev and Iudin were included in the ten-member commission, suggesting the heights to which Soviet scholars had climbed within the Party. Making the point explicit, the resolution stated that "Comrade Stalin's work *Economic Problems of Socialism in the USSR* will guide the fundamental tenets of the revision of the Program."[129] The political impact of Stalin's short book was muted, however. In the following months the revised Program never materialized, and after Stalin's death in March 1953 the effort was abandoned.

The book's impact on Soviet academia and education was more impressive. Within a month of the publication of *Economic Problems*, the Central Committee had organized a campaign on its behalf. Following the Party's lead, the Academy of Sciences and Ministry of Education revised their scholarly agendas in light of Stalin's work, organized conferences addressing it, published editorials in all their major journals praising it, and oversaw dissertations exploring its meaning.[130] *Economic Problems* became the focus of ubiquitous praise in the press and in scholarly meetings for months.

Needless to say, Stalin's work was also a guiding force for the economists responsible for the completion of the textbook. Throughout the fall the authors continued to work diligently on finishing the book. Fortunately for them, they had been kept abreast of Stalin's writings on political economy, and *Economic Problems* did not come as a surprise. Even Stalin's death in March 1953 does not seem to have slowed them down. At the end of March Ostrovitianov reported to Khrushchev that the commission's work on the textbook was near-ing completion. They had missed their March 1 deadline, but they promised a finished product no later than May 15, 1953. Agitprop suggested that at that point the authors' work was done and that they could finally return to their primary responsibilities.[131] But upon further review the Central Committee called for additional revisions. The textbook was finally published in 1954, eighteen months after Stalin's death and seventeen years after the project had been initiated.[132] The *Textbook on Political Economy* did not share the aura of infallibility that surrounded its fraternal twin, the *Short Course* on the history of the Party. Like the sickly and slow-developing runt of a litter, the political economy textbook was nurtured so persistently by Stalin and others in the Central Committee that it barely had the opportunity to stand on its own. It was released into a world of political uncertainty after Stalin's death. The change in atmosphere from 1938, when the *Short Course* was published, to 1954, when the *Textbook on Political Economy* appeared, was apparent in the book's presentation. There was no stamp on its cover declaring the book "Edited by a Commission of the Central Committee" or "Authorized by the Central Committee." The authors shied away from presenting their work as definitive, stating that they intended "to continue to work on the further improvement of the text, on the basis of critical observations and suggestions which readers will make when they have acquainted themselves with the first edition."[133] The discussion would continue.

CHAPTER 8

CONCLUSION

Science and the Fate of the Soviet System

Joseph Stalin reigned over one of the most ambitious social revolutions ever undertaken; the goal was to end meaningful conflict and create the "kingdom of freedom" on earth. Stalinists were bound by ironclad laws (the victory of communism was inevitable) and dedicated to action (the Party, state, and individuals had to struggle to bring about that victory.) Soviet socialism was both a rigorous scientific theory and a living, creative practice. And it was all-encompassing: from the "free will of electrons" to the origins of languages, from the idealist germ plasm to the materialist conditioning of the human brain, there was not a subject in the universe beyond the reach of Party doctrine.

In the late 1940s and early 1950s six academic disciplines became the focus of Party-sponsored debate. Officially, these debates successfully reconciled Marxist-Leninist ideology and scientific thought. Delivering the major policy speech at the XIX Party Congress in 1952, Georgii Malenkov lauded the value of the Party's interventions in science: "The well-known discussions in philosophy, biology, physiology, linguistics and political economy have unveiled serious ideological digressions in various fields of science, have stimulated and developed criticism and controversies and have played an important role in the advancement of science."[1] Malenkov and other Party leaders insisted that Stalin's guidance had helped end ideological confusion in Soviet scholarship.

Scientists and academics knew better: applying newly established truths to concrete subjects proved difficult. The discussions had been intended to clarify the Party's position on scientific controversies, but instead they exacerbated the ideological crisis. The problems were particularly vexing for scholars contributing to the *Great Soviet Encyclopedia*, the official repository for all Soviet knowledge. In 1951 the encyclopedia's chief editor, A. A. Zvorykin, directly confronted the ambiguities of the scientific discussions in a long memo to Malenkov. He noted that Stalin's articles on linguistics had suggested that some categories were neither part of the base nor part of the superstructure and were therefore beyond ideology. This made it problematic, if not impossible, to write acceptable encyclopedia entries for such basic terms as "science," "natural science," and "social science." Zvorykin's contributors also hesitated to address subjects in physics, because they were unsure whether there was a distinction between philosophical and physical understandings of matter. Unsettled disputes in political economy left the entries for "money," "production," and the "law of value" up for discussion. While accepting that the Agricultural Academy discussion and the Pavlov Session "successfully pushed work forward," Zvorykin did not know how to address the subject of "Darwinism" and noted that while general descriptions of Pavlov's work were possible, concrete ideas about Pavlov and medicine had not yet been formulated.[2]

Party functionaries in the Science Section responded with dubiety. On the one hand, they reported to Malenkov that Zvorykin's questions had been answered over the last few years in the "scientific discussions in biology, philosophy, linguistics, and physiology" and that the political economy questions would be answered in the forthcoming textbook. (Obviously they could not know that the book was still three years from publication.) On the other hand, they admitted that a "large number of scientific problems require further elaboration" and conceded that Zvorykin should publish entries that took into account the current state of scientific knowledge, without attempting to "prematurely resolve unstudied, controversial, or unclear questions."[3] After six discussions dedicated to defining the relationship between ideology and knowledge, bureaucrats surrendered ground, admitting that scientific ideas could be valid even if their exact relationship to Marxism-Leninism had not been worked out.

Even with postwar Stalinism's extremely limited range of discourse, scholars in each academic field formulated and defended conflicting views about what constituted Soviet science. Each discussion included a painful, dangerous, and sometimes imaginative search for acceptable ways of presenting a worldview that was compatible with both canonical texts and new scientific discoveries. Nothing in Marx or Lenin predetermined that genetics would be outlawed, that Ivan Pavlov would be knighted as a Soviet hero, that Nikolai Marr would be first praised and then lambasted, or that Georgii Aleksandrov's prizewinning book would become the symbol of all that was wrong with Soviet philosophy. Yet, the Central Committee's decisions were neither wholly arbitrary nor independent of the scholars involved. Decisions in each discussion influenced subsequent ones. Though in each case the combination of events and actors was complex, certain themes emerge. High politics, patronage networks, Cold War imperatives, and advice that flowed upward from the Science Section and from scientists themselves all influenced Stalin's decisions and the discussions' outcomes.

Domestic high politics played a crucial role in shaping the discussions. In 1946 and 1947 Stalin made it clear to Andrei Zhdanov that Zhdanov had to rectify "shortcomings" on the "ideological front" if he wanted to remain in favor. By turning his critical attention to problems with *History of Western European Philosophy*, Stalin invited others to attack Aleksandrov's handling of the ideological front more generally. The resulting philosophy discussion led to a shake-up of Agitprop, leaving Zhdanov significantly weaker in the Party. Malenkov did not hesitate to take advantage of the confusion within Zhdanov's bailiwick. When Agitprop and the Science Section consistently rebuffed Lysenko's attempts to denounce his scientific detractors on ideological grounds, Malenkov supported him by steering his memos and complaints to Stalin's desk. Even when infighting among Stalin's lieutenants was not a central factor, political patronage helped shape the outcome of scientific discussions. Beria, who had little personal concern for ideological questions, used his political clout to assist his clients. He helped bring Chikobava's criticism of Marr's linguistic theories to Stalin's attention. He also defended physicists who worked under him on the Soviet atomic bomb project from ideological attacks.

The scientific discussions were about more than political maneuvers at the highest ranks of the Party and state. They also display the difficulties inherent in integrating Soviet "patriotism" with Marxist-Leninist ideology. In each discussion some scholars resisted Russocentrism and, taking their clues from the early 1930s, emphasized the universal foundations of Marxism and the centrality of class—as opposed to nationality—in determining the contours of Soviet science. But in the 1940s and early 1950s they consistently found themselves reprimanded by the Party. In the philosophy discussion Aleksandrov faced criticism for overemphasizing the role of Hegel and underemphasizing the role of Russians in the history of Marxism. In biology, Lysenko gained an advantage over his "Mendelian-Morganist" detractors by emphasizing that his ideas were homegrown and based on the specifically Russian scientific tradition of Michurin. The press touted Pavlov as a Russian scientist. The linguistics discussion led to greater emphasis on Russian language and literature in universities and research institutes. Stalin implored the authors of the political economy textbook to pay more attention to Russian examples in their depiction of the history of capitalism.

The centrality of Russia and Russians in the history of all thought brought with it disdain for all things "foreign" or "cosmopolitan." In physiology, the primary "losers" were Orbeli (an Armenian), Beritashvili (a Georgian), and

Shtern (a Jew), while the party supported the Russians Bykov and Ivanov-Smolenskii. The Central Committee fretted over the high percentage of Jews in physics and economics. Jewish physicists were—to a great extent—protected by an "atomic shield": Landau, Al'tshuler, and other Jewish physicists thrived professionally during the late 1940s. "Cosmopolitanism," however, cost Abram Ioffe his job. Economics was hit harder. But Jews such as Varga, Trakhtenberg, and Rubinshtein remained active participants in scholarly debates, despite obvious setbacks during the anti-Semitic fervor of the late-Stalin period. Leont'ev continued to meet with Stalin and work on the political economy textbook, although other economists, such as Ostrovitianov, clearly benefited from being ethnic Russians. In each discussion participants struggled to determine how "cosmopolitanism" and Russocentrism affected their fields.

The discussions also reflected the USSR's international aspirations. Cold War tensions between the United States and the USSR contributed to the urgency with which Stalin and the Communist Party asserted that Marxism-Leninism made possible discoveries that were not available to Western science. That said, there is no extant evidence that Stalin coordinated the scientific discussions with any specific diplomatic or military initiatives of the early postwar period.[4] Instead, he saw science as a sphere of Cold War competition in its own right. Whichever side produced the most advanced science would have a rhetorical advantage on the "ideological front." Within this context Stalin sought to show how Soviet science was both different from and superior to Western science. The Cold War was a competition between opposing ideologies—not just between opposing economies and militaries.[5] The role that Stalin had assigned scientists in the late 1940s only grew more important after Stalin's death as the arms race, the space race, and competition for Nobel Prizes in science heated up. Both sides understood scientific breakthroughs as standards by which to judge progress; both accepted a quintessentially modern worldview rooted in science and rationality.

The scientific discussions suggested that ideology was malleable. When a letter to the Central Committee pointed out the contradictions between what Stalin had written in the 1930s and what he wrote in 1950, Stalin dismissed the charge with the observation that only a "Talmudist" would insist that theories developed in one period of historical development were necessarily correct during all other periods. Of course, not every ideological tenet was subject to reexamination. In some sense everybody "knew" and therefore did not question that single-party rule, heavy industry, and a planned economy were essential for building socialism. But the scientific discussions showed that ideology was not entirely rigid. If it had been, Aleksandrov would never have written his "Eurocentric" history of philsophy, Yuri Zhdanov would have never attacked Lysenko, a physics conference would have gone off smoothly, the Central Committee would not have supported Marr so actively, and the political economy textbook would not have been so difficult to write. At various times Andrei Zhdanov, Malenkov, and other Party secretaries, the head of Agitprop, the head of the Science Section, the president of the Academy of Sciences, the minister of education, and countless scientists in every discipline miscalculated what ideologically correct science was supposed to look like. Stalin was the only person who could keep up with his own evolving interpretations.

Over the last years of his life Stalin came to believe in the "objective" truth of science and in the role of discussion in arriving at that truth. Now that the archives are open, a pattern to Stalin's interventions emerges that was not always visible to his subordinates at the time. In each case Stalin addressed the relationship between scientific truth and Party-dictated truth. The philosophy discussion provides a starting point for understanding the development of Stalin's ideas. In 1946 he had criticized Aleksandrov's book for being "too objective" when the "ideological front" called for politically engaged work. Philosophers debated whether being a Marxist-Leninist meant discovering objective laws about society or meant trumpeting the infallibility of the Party and the superiority of the Soviet system. The official answer, which was worked out in the course of the 1947 discussion, was that they were supposed to do both. But tensions between philosophers as scientists and philosophers as publicists remained visible in the pages of *Questions of Philosophy*, where some wrote "academic" articles and others called for greater engagement on the "ideological front."

The shift in Stalin's views about science began as early as 1948, with his editing of Lysenko's infamous speech. Not only did the "coryphaeus" delete Lysenko's references to class-based science, he confidently addressed the scientific issues at stake. When he decided to throw the Party's support behind Lysenko, he chose a subtle approach. Biologists gathered at a scientific meeting where the Party's position was not clear to the participants. Even when Stalin allowed Lysenko to confront his challengers by revealing that the Party backed him, Stalin kept his personal role hidden. At this point, he recognized that the Party could support certain scientific theories but that appearing to dictate them would only diminish their prestige.

Other members of the Party and state elite also tried to draw a line between science and ideology. In preparation for the aborted physics conference, the minister of education, Sergei Kaftanov, insisted that there was a distinction between the validity of a scientific concept (which was a matter for scientists to decide) and its philosophical implications (which was a matter for philosophers and the Party.) Stalin pushed the issue further, and in a public forum, when his essays on linguistics rejected the notion that all thought was somehow part of either the economic base or the ideological superstructure. He also attacked scientific monopolies and insisted that science could flourish only in an atmosphere of open discussion. He elaborated these contentions in his writings on political economy, declaring that scientific laws were "objective" and universal. Again, he determined that a discussion in which people "argued and tested one another" was the best format for working out the details. Again and again in meetings at the Kremlin, Stalin insisted that political economists were supposed to be scientists, not apologists for the Party's policies.

How seriously can we take Stalin's proclamations on behalf of "objective" science and "free and open discussions"? Some historians have questioned whether Stalin wrote his essays at all. That can now be settled definitively.[6] The archives are full of drafts of essays, notes, and editorial comments, all in Stalin's handwriting. Others have suggested that Stalin's articles were motivated by political goals and therefore their content is practically irrelevant.[7] While political infighting and personal vendettas played a role in some battles, Stalin's persistent concern for the details of scholarship suggests that there was more at stake than power politics. Why else would Stalin depend on elaborate, time-consuming, and sometimes unpredictable scientific discussions rather than simply dictate policy? He took himself seriously as a thinker and was clearly pleased that others took him seriously as well. Long after the institutional and personal issues had been settled, he read the countless letters addressed to him and the Central Committee asking for clarifications and elaborations of his ideas. His interest in science was too thorough and consistent across time to be about politics alone. Newly accessible documents make this all the more clear. They reveal a man more engaged in the substance of the discussions in private than he had appeared to be in public. The issues mattered to Stalin.

Stalin may have had personal motivations for participating in scholarly disputes. The biographer Robert Tucker has described Stalin's 1925 book *Foundations of Leninism* as an effective means for him to "prove himself a Bolshevik leader of large theoretical horizons." In the struggle for succession, the book helped Stalin shore up his weak credentials as a Marxist-Leninist thinker, while outlining a version of Leninism that was both in line with his doctrinaire notions of ideology and accessible to the new, young, less intellectual Party cadres who would help him secure power.[8] His postwar interventions in science display that same urgency to become a great theorist in the tradition of Marx, Engels, and Lenin. He identified himself as a scholar and saw that identity as a central component of being a successful Marxist-Leninist leader.[9]

Elements of Stalin's postwar interventions were intended to invigorate, not destroy, Soviet science. His forays into scientific debate revealed a remarkable intellectual arrogance (he regularly schooled people in fields of their own expertise) and the crude adaptation of political methods to settle scientific debates (he developed strategies for how best to make sure the side he supported emerged from discussions "victorious"). But his calls for "a battle of opinions" and "freedom of criticism" had far-ranging effects on Soviet science. Heeding Stalin's call for the end of "monopolies" in various fields of scholarship, Party bureaucrats and scientists reevaluated the relationship between science and ideology. In the early 1950s Central Committee leaders collected and distributed letters from scientists who were critical of Lysenko. In the summer of 1952—that is, less than four years after Stalin gave his approval to Lysenko to purge biology of his enemies—members of the Politburo outlined plans to change the Party's stand on Lysenko and his theories. In challenging Lysenko, they did not offer doctrinal arguments or assert the Party's prerogative to make scientific decisions. Instead, they decided that Lysenko's activities were hindering scientific progress, despite their ideological merits.[10]

So long as Stalin remained alive, the potential for science to conflict with ideology was limited. Confident in the ultimate validity of Marxism-Leninism, Stalin encouraged scientists to contribute to the advancement of ideology, rather than simply to reflect it. Scientists in a full range of fields took advantage of the room to maneuver. In a letter to Stalin that referred directly to the leader's article on linguistics and essay on political economy, the physicist Peter Kapitsa applauded the idea that the laws of nature were the same everywhere, "on the equator and the poles, for us and for capitalists and for cannibals." Stalin's essays had helped doctrine reach the point that the atomic bombing of Hiroshima and Nagasaki had made clear in practice: Marxism-Leninism did not have a monopoly on the discovery and productive application of scientific laws. With science liberated (at least partly) from doctrinal concerns, Kapitsa suggested that Soviet scientists should be given the freedom and independence to uncover universal laws, which, in turn, could be used to help build communism in the USSR. Though it is not clear whether Stalin read this letter, in 1954 his successor, Nikita Khrushchev, used it as a basis for a discussion at the presidium of the Central Committee on the organization of Soviet science.[11]

Under Khrushchev scientists became elite members of Soviet society whose status derived from their ability to innovate, not from their ability to repeat Party slogans. The Party allowed scientists greater autonomy over their own fields.[12] The 1961 Party Program stated that "the Party will do everything to enhance the role of science in building communist society."[13] As the scholars Matthew Evangelista and David Holloway have shown, some physicists transformed their valuable scientific know-how into political influence with leaders at the top of the Soviet bureaucracy.[14] Physicists were not alone. In the middle of the 1950s the Party approved the reestablishment of Evgeny Varga's institute under the new title "Institute of the World Economy and International Relations." It became a crucial center for training a younger generation of scholars whose understanding of Western economics and politics was less encumbered by official ideology. Veterans of the Stalin-era debates about the relationship between doctrine and economic analysis taught a new generation about the dangers of a calcified ideology. Varga, who during Stalin's life had not been shy about expressing ideas that challenged the accepted understanding of Marxism-Leninism, took the lead and encouraged his disciples to follow. Georgii Arbatov—a key player in bringing about Mikhail Grobachev's reforms—described Varga's institute in the late 1950s as an " 'incubator' for a new generation of international economists . . . and foreign policy experts . . . [where] dogmas about capitalist stagnation, total impoverishment of the Western working class, and others were rejected while new concepts came into political circulation [such as] European integration . . . multiple paths to third-world development . . . and so on."[15] Arbatov did not know that some of those same topics had also been raised at the 1951 political economy discussion.

After Stalin's death scholars felt more confident developing ideas that may or may not have vindicated official Party positions. Most scientists, of course, shied away from publishing direct challenges to the ideology. But without a coryphaeus of science, Party leaders also appeared reluctant to push for harmony between Party truths and scientific truths anywhere but on the surface. The use of ornamental quotations from Marx, Engels, and Lenin at the front of serious and innovative scientific papers buttressed the notion that ideology and science remained united at least in principle. In practice, few people challenged the relatively broad parameters the Party set for scholarship in the post-Stalin period. Among the few who did was the physicist Andrei Sakharov, who began as a prominent scientific insider, offering advice to political leaders on technical issues. In the 1960s and 1970s he and his fellow dissidents from the scholarly community became vocal about the value of science and rationality for addressing political problems.[16] When the Party rebuffed them, they challenged the Party's claim to a scientific worldview. In his acceptance speech for the Nobel Peace Prize in 1975 Sakharov argued that "intellectual factors play a special role in the mechanism of progress" but that in the socialist countries the "vulgar, ideological dogmas of official philosophy" had distorted the process. "Progress is possible and innocuous," he continued, "only when it is subject to the control of reason."[17] Elsewhere, he wrote of his anxiety that "the scientific method of directing policy, the economy, arts, education and military affairs still has not become a reality."[18] In 1950 and 1951, Stalin had insisted that Marxist-Leninism had to be scientific. Sakharov brought this notion to its logical conclusion by arguing that when reason (as understood by scientists) clashed with official ideology, it was the doctrine that had to give ground, even on moral and political questions.

When Gorbachev came to power in the 1980s, he believed that a full-scale reconciliation between ideology and scholarship would reinvigorate socialism. Ideology had suffered from calcification in the years of stagnation, while science had grown confident after years of government support. Perhaps a jolt of rational thinking could save Marxism-Leninism. Using words that, ironically, echoed Stalin, Gorbachev chided experts for their "inability to tell the truth" and implored top economists to stop the "false idealization of reality."[19] During glasnost a wide range of scholars took his call for openness seriously. But rather than reinforcing the Party's claim to legitimacy, as Gorbachev had hoped, they unleashed a torrent of truths that helped wash away the ideological foundations of the system.[20]

Soviet attempts to clarify the relationship between ideology and science can be understood within the context of similar efforts in other modern states. Though Stalin's interventions in science were unprecedented in the history of the twentieth century, his contemporaries also contemplated the connection between science and their political ideologies. In Nazi Germany, Adolf Hitler turned to biological theories to support his racialist thinking, but he remained much more interested in astrology, superstition, and "will" than in reason.[21] Even when a 1935 propaganda poster declared Hitler a "Doctor For the German People," his expertise was never considered all-encompassing. Unlike the Communist Party under Stalin, the Nazi Party never set for itself the goal of showing the compatibility of its ideology with all fields of knowledge.[22] In the postwar period, German scientists confronting the legacy of their involvement in the Third Reich debated whether their failure to confront Nazi ideology had been caused by their lack of political engagement or by their all-too-easy acceptance of the politicization of science.[23] But in either case, unlike Soviet scientists after the war, they remained reluctant to have their work too closely associated with the goals of the state.

Communist China has confronted some of the same questions about ideology and science that arose in the USSR. Not surprisingly, Mao Zedong and other Chinese Communist Party officials came up with a familiar range of responses. During the Hundred Flowers campaign they conceded that the "natural sciences . . . have no class character" and that the Party should admit its "total ignorance" about matters of science.[24] Then, the Cultural Revolution witnessed the Chinese Communists Party's ruthless repression of scientists and its assertion of ideological control over research. In the post-Mao period, a small but significant group of Chinese scientists followed the lead of their counterparts in the USSR and used their professional prestige to put pressure on the Party to create a more liberal and rational system.[25]

The USSR's Cold War adversary, the United States, provides one of the most illuminating comparisons with Soviet attempts to work out the relationship between ideology and science. During the Second World War President Franklin Delano Roosevelt respected the judgment and advice of scientists, even if he closely guarded policy decisions regarding their findings. After the war, the American scientific community adopted the language of "free enterprise" to argue that science worked best when its practitioners were exempt from the scrutiny and demands of the politicians who funded their increasingly expensive projects. They defended their autonomy by self-consciously contrasting their own democratic scientific ethos with Lysenkoism and the situation in Soviet genetics.[26]

During the 1960s and 1970s, U.S. and Soviet ideas about science and politics were heading in opposite directions. As Soviet ideology followed Stalin's lead and increasingly conceived of science as "objective," American historians and sociologists came to understand science as intertwined with politics and society and therefore subject to the same types of criticisms and analyses as other human endeavors. Today, many American scientists continue to defend a strict line between scientific knowledge and politics or ideology. Indeed, according to this way of thinking, their value to the state and society comes precisely from their objectivity. They are grateful for continued financial support from the state but adamant about maintaining a system that allows scientists themselves to determine what constitutes correct or useful science.[27] During the 1980s Soviet scientists hoped to follow the example of their American colleagues, who managed (for the most part) to have both funding and autonomy. Instead they were granted more control over their disciplines, but less money. The economic collapse of the 1990s led scientists to emigrate in search of support for their research or to head for more lucrative professions in the former USSR. Those who remained in science showed a marked nostalgia for the Soviet system and even aspects of the Party's ideology, which had emphasized the importance of science for society and in turn funded science aggressively. Despite decades of political and ideological interference by the Party, many ex-Soviet scientists would rather have funding without political freedom than political freedom without funding.[28]

Ultimately, comparisons with other governments' attempts to establish a working relationship between politics and science serve only to highlight the peculiarity of Stalin's stint as a comrade scientist. A range of Soviet and foreign leaders in the twentieth century recognized the importance of science in underpinning the ideological coherence of their respective political systems. But only Stalin claimed to be able to unify power and knowledge. The years 1947 to 1949 represented a high-water mark in Stalin's drive to integrate all thought into Soviet ideology. By 1950 he began to retreat from his most ambitious definition of what Marxism-Leninism could mean. It is, of course, ironic that he would decree from on high the benefits of free and open discussions. But his statements on behalf of "objective" knowledge represent the tentative and awkward first steps toward accepting science as a subject beyond the Party's ideological reach. As a result, the seeds of science's rising prestige in the Soviet Union in the post-Stalin decades were planted during the twilight of Stalin's reign. The process could get under way in earnest only after Stalin's death, when the personal embodiment of the union of ideology and science disappeared. Subsequent Soviet leaders did not share Stalin's compulsion to be an expert in all things. Khrushchev, Brezhnev, and Gorbachev did not assume the label "coryphaeus of science." The Party continued to control decisions about the direction of scientific research and occasionally intervened in scientific disputes, but never with the energy or consistency with which it had done so through the 1940s. By backing away from the sweeping and aggressive claims of that era, it could reestablish its legitimacy on more solid ground. Faith in the objectivity and universality of science became more pronounced. Scientists took advantage of this trend and carved out for themselves "little corners of freedom" and "islands of intellectual autonomy" isolated from Party decrees.[29] That these safe havens germinated during Stalin's time is not easy to reconcile with the brutality and capriciousness of his regime. But even as Stalin dictated fundamental truths, he gradually came to accept scientists' authority to ascertain laws that were beyond human ability to create or control. It was only a matter of time before some scientists used this privileged access to truth to challenge the rationality of the ideology and thus the legitimacy of the regime. A socialism that was not scientific was no socialism at all.

NOTES

Archival Collections Cited

Arkhiv Rossiiskoi Akademii Nauk—**ARAN** (Archive of the Russian Academy of Sciences)

    fond 2 (presidium of the Academy of Sciences)
    fond 532 (Physics Institute of the Academy of Sciences)
    fond 578 (State Commission for Celebrating Pavlov's 100th Birthday)
    fond 615 (Scientific Secretariat of the Academy of Sciences)
    fond 678 (Institute of Language and Thought of the Academy of Sciences)
    fond 684 (G. F. Aleksandrov)
    fond 596 (S. I. Vavilov)
    fond 1513 (E. S. Varga)
    fond 1515 (A. A. Maksimov)
    fond 1602 (V. V. Vinogradov)
    fond 1636 (P. F. Iudin)
    fond 1705 (K. V. Ostrovitianov)
    fond 1877 (Institute of Economics of the Academy of Sciences)
    fond 1922 (Institute of Philosophy of the Academy of Sciences)

Gosudarstvennyi arkhiv Rossiiskoi Federatsii—**GARF** (State Archive of the Russian Federation)

    fond 9396 (Ministry of Higher Education)
    fond 5446 (Council of Ministers)

Rossiiskii gosudarstvennyi arkhiv ekonomiki—**RGAE** (Russian State Economic Archive)

    fond 8390 (Lenin All-Union Agricultural Academy—VASKhNIL)

Rossiiskii gosudarstvennyi arkhiv sotsial'no-politicheskoi istorii—**RGASPI** (Russian State Archive for Socio-Political History—Former Central Party Archive) fond 17 (Central Committee of the All-Union Communist Party [Bolsheviks])

op. 3 (Politburo protocols)
op. 115–121 (Orgburo and Secretariat protocols and materials)
op. 125 (Agitation and Propaganda Administration, 1939–1948)
op. 132 (Section for Agitation and Propaganda, 1948–1956)
op. 133 (Sections for Science, Arts, and Education, 1951–1953)
op. 138 (Agriculture Section)
op. 165 (materials for Central Committee conferences)

fond 77 (A. A. Zhdanov)
fond 83 (G. M. Malenkov)
fond 88 (A. S. Shcherbakov)
fond 558 (I. V. Stalin)
fond 592 (XIX Party Congress)
fond 629 (P. N. Pospelov)

Tsentral'nyi Gosudarstvennyi arkhiv obshchestvennykh dvishenii goroda Moskvy—TsAODM (Central Moscow Archive of Social Movements—Former Moscow Party Archive)

fond 2501 (Institute of Philosophy of the Academy of Sciences)
fond 3746 (Institute of Physiology of the Academy of Medical Sciences)
fond 992 (presidium of the Academy of Sciences)
fond 2505 (Institute of Linguistics of the Academy of Sciences)
fond 6862 (Physics Institute of the Academy of Sciences)

Soviet Periodicals Cited
Bol'shevik
Botanicheskii zhurnal
Izvestiia
Izvestiia Akademii Nauk SSSR
Komsomol'skaia pravda
Krasnyi flot
Kul'tura i zhizn'
Literaturnaia gazeta
Meditsinskaia gazeta
Novoe vremia
Partiinaia zhizn'
Pod znamenem marksizma
Pravda
Priroda
Sovetskaia kniga
Trud
Vestnik Akademii Nauk
Voprosy ekonomiki
Voprosy filosofii

Voprosy istorii
Zvezda

CHAPTER 1
INTRODUCTION

1. The phrase was invoked in 1939 when Stalin became a member of the Soviet Academy of Sciences, but it did not come into widespread use until after the war.

2. The Russian word for science is *nauka* and, like the German word *Wissenschaft*, it connotes the pursuit of knowledge in general and is not restricted to the investigation of natural phenomena. At times I use the word "scholarship" and "scholars" instead of "science" and "scientists," in the hope that it will convey the broader sense of the word *nauka*. It is often the case that problems with translation reveal issues of deeper significance, and this is no exception. In the Soviet Union barriers between the natural, human, and social sciences were blurred.

3. The term "discussion" requires further explanation. As Alexei Kojevnikov has spelled out, Soviet scientists and Central Committee bureaucrats used the term *diskussiia* [discussion or disputation] to refer to a specific forum where "temporary, public disagreements over important political questions" were allowed. Sometimes these "discussions," however, were also called "scientific sessions" or "conferences." Borrowing from the usage of the time, I use "discussion" to refer to the process of scholarly debate and to the specific sessions and conferences where disputes were aired. See Alexei Kojevnikov, "Rituals of Stalinist Culture at Work: Science and the Games of Intraparty Democracy circa 1948," *Russian Review* 1998 (January): 33.

4. Recognizing the ideological relevance of each of these disciplines, however, does not necessarily mean that the Central Committee set out in 1946 to hold discussions in these fields as part of a single, predetermined campaign. To the contrary, methodological, institutional, practical, and personal idiosyncrasies in each discipline affected the discussions' planning, content, and impact. It seems somewhat arbitrary that Stalin did not end up adjudicating disputes about history, chemistry, and mathematics. For an excellent analysis of the shifting boundaries between ideology and science in the USSR during Stalin's time and after, see Slava Gerovitch, *From Newspeak to Cyberspeak: A History of Soviet Cybernetics* (Cambridge, 2002).

5. Recent scholarship has also emphasized the influence of ideology, in combination with other factors, on Soviet foreign policy during the Cold War. See Vladislav Zubok and Constantine Pleshakov, *Inside the Kremlin's Cold War: From Stalin to Khrushchev* (Cambridge, 1996), 1–8; Nigel Gould-Davies, "Rethinking the Role of Ideology in International Politics during the Cold War," *Journal of Cold War Studies* 1 (1999): 90–109; and Melvyn P. Leffler, "The Cold War: What Do 'We Now Know'?" *American Historical Review* 104 (1999): 501–524.

6. On this point, see Loren Graham, *The Soviet Academy of Sciences and the Communist Party, 1927–1932* (Princeton, 1967), 32–33.

7. Loren Graham, *Science in Russia and the Soviet Union* (Cambridge, 1993), 88–90.

8. Sheila Fitzpatrick, ed., *Cultural Revolution in Russia, 1928–1931* (Bloomington, 1984).

9. David Joravsky, *Soviet Marxism and Natural Science* (New York, 1961).

10. Alexei Kojevnikov, "Dialogues about Knowledge and Power in Totalitarian Political Culture," *Historical Studies in the Physical and Biological Sciences* 30, 1 (1999): 234–239.

11. Nikolai Krementsov, *Stalinist Science* (Princeton, 1997), 116.

12. William Taubman, *Stalin's American Policy: From Entente to Détente to Cold War* (New York, 1982), 133–134, and Melvyn P. Leffler, *A Preponderance of Power: National Security, the Truman Administration, and the Cold War* (Stanford, 1992), 102–104.

13. J. Stalin, *Speeches Delivered by J. V. Stalin at a Meeting of Voters of the Stalin Electoral District, Moscow* (Moscow, 1950), 41. See also Zubok and Pleshakov, *Inside*, 35.

14. Quoted in David Holloway, *Stalin and the Bomb* (New Haven, 1994), 148.

15. For more on the deal between the regime and various groups of elites, see Vera Dunham, *In Stalin's Time: Middleclass Values in Soviet Fiction* (Durham, 1990).

16. Ethan Pollock, "Conversations with Stalin on Questions of Political Economy," Cold War International History Project (Working Paper No. 33), 2001, 35.

17. Elena Zubkova, *Russia after the War: Hopes, Illusions, and Disappointments, 1945–1957* (New York, 1998); Donald Filtzer, *Soviet Workers and Late Stalinism: Labour and the Restoration of the Stalinist System after World War II* (Cambridge, 2002); Chris Burton, "Medical Welfare during Late Stalinism: A Study of Doctors and the Soviet Health System, 1945–1953," Ph.D. dissertation, University of Chicago, 2000.

18. Zubok and Pleshakov, *Inside*, 110–137.

19. Quoted in Holloway, *Stalin and the Bomb*, 207.

20. Nikolai Krementsov, *The Cure: A Story of Cancer and Politics from the Annals of the Cold War* (Chicago, 2002), 126–128.

21. Yuri Slezkine, "The USSR as a Communal Apartment, or How a Socialist State Promoted Ethnic Particularism," *Slavic Review* 53 (Summer 1994): 414–452; Terry Martin, *The Affirmative Action Empire: Nations and Nationalism in the Soviet Union, 1923–1939* (Ithaca, 2001); and David Brandenberger, *National Bolshevism: Stalinist Mass Culture and the Formation of Russian National Identity, 1931–1956* (Cambridge, 2002).

22. For more on postwar anti-Semitism, see Yuri Slezkine, *The Jewish Century* (Princeton, 2004), and G. V. Kostyrchenko, *Tainaia politika Stalina: vlast' i antisemitizm* (Moscow, 2001).

23. To be precise, the Agitation and Propaganda Administration (*upravlenie*) was renamed the Section (*otdel*) for Propaganda and Agitation in the Central Committee reorganization of 1948. I refer to both as Agitprop. Similarly, the Science Section went through various transformations in the period. In 1950 it became nominally independent of Agitprop (although the personnel continued to work together). In the fall of 1952 it was broken down into three sections, one for natural and technical sciences, one for philosophical and legal sciences, and one for economic and historical sciences. A reunited Science and Culture Section was created in March 1953. For simplicity's sake, throughout the book I use the shorthand "Agitprop" and "Science Section."

24. Stephen Jay Gould, *Hen's Teeth and Horse's Toes* (New York, 1983), 135.

25. In addition to the various works cited in this introduction, recent scholarship that addresses the postwar period includes Yoram Gorlizki and Oleg Khlevniuk, *Cold Peace: Stalin and the Soviet Ruling Circle, 1945–1953* (Oxford, 2004); Julie Hessler, *A Social History of Soviet Trade: Trade Policy, Retail Practices and Consumption, 1917–1953*

(Princeton, 2004); Thomas Lahusen, *How Life Writes the Book* (Ithaca, 1997); Douglas R. Weiner, *A Little Corner of Freedom: Russian Nature Protection from Stalin to Gorbachev* (Berkeley, 1999).

26. Kendall E. Bailes, *Technology and Society under Lenin and Stalin* (Princeton, 1978); Loren Graham, *Science Philosophy and the Soviet Union* (New York, 1972); David Joravsky, *The Lysenko Affair* (Cambridge, 1970); Joravsky, *Soviet Marxism and Natural Science* (Columbia, 1961); and Joravsky, *Russian Psychology: A Critical History* (Oxford, 1989); Paul Josephson, *Physics and Politics in Revolutionary Russia* (Berkeley, 1991); Linda L. Lubrano and Susan Gross Solomon, eds., *The Social Context of Soviet Science* (Boulder, 1980); Alexander Vucinich, *Empire of Knowledge: The Academy of Sciences of the USSR (1917–1970)* (Berkeley, 1984). Although not a history of science, Werner Hahn's *Postwar Soviet Politics: The Fall of Zhdanov and the Defeat of Moderation, 1946–1953* (Ithaca, 1982) provided essential details culled from published materials about the Central Committee apparatus in the late Stalin period.

27. Joravsky, *The Lysenko Affair*, xi.

28. As recent access to state, Party, and academy archives has grown, a new generation of scholars has begun to pose new questions about what happened in particular academic disciplines during the postwar period. In each of the following chapters I provide detailed references to this work.

29. David A. Hollinger, "The Defense of Democracy and Rober K. Merton's Formulation of the Scientific Ethos," *Knowledge and Society* 4 (1983): 1–15, and Jessica Wang, "Merton's Shadow: Perspectives on Science and Democracy since 1940," *Historical Studies in the Physical and Biological Sciences* 30 (1999): 279–306.

30. Quoted in Peter Novick, *That Noble Dream: The "Objectivity Question" and the American Historical Profession* (Cambridge, 1988), 296–297. To approximate the official Soviet view in the same year, simply interchange the words "American" and "Soviet Union" in Conant's statement.

31. Quoted in A. S. Sonin, "Soveshchanie, kotoroe ne sostoialos'," *Priroda* 1990 (4): 97.

CHAPTER 2
"A MARXIST SHOULD NOT WRITE LIKE THAT"

1. On Soviet society and economy after the war see Elena Zubkova, *Russia after the War: Hopes, Illusions, and Disappointments, 1945–1957* (New York, 1998), and Donald Filtzer, *Soviet Workers and Late Stalinism: Labour and the Restoration of the Stalinist System after World War II* (Cambridge, 2002). On the early Cold War from the Soviet perspective, see Vladislav Zubok and Constantine Pleshakov, *Inside the Kremlin's Cold War: From Stalin to Khrushchev* (Cambridge, 1996).

2. Loren R. Graham, *Science in Russia and the Soviet Union: A Short History* (Cambridge, 1993), 99–103.

3. David Brandenberger, *National Bolshevism: Stalinist Mass Culture and the Formation of Modern Russian National Identity, 1931–1956* (Cambridge, 2002).

4. For more on the Institute of Red Professors during the 1920s, see Michael David-Fox, *Revolution of the Mind: Higher Learning among the Bolsheviks, 1918–1929* (Ithaca, 1997).

5. Rossiiskii gosudarstvennyi arkhiv sotsial'no-politicheskoi istorii (RGASPI), f. 83, op. 1, d. 7, ll. 107–107b. G. S. Batygin and I. F. Deviatko, "Delo akademika G. F. Aleksandrova: epizody 40-x godov," *Chelovek* 1993 (1), 135; Iu. I. Krivonosov, "Srazhenie na filosofskom fronte: filosofskaia diskussiia 1947 goda—prolog ideologicheskogo pograma nauki," *Voprosy istorii estestvoznaniia i tekhnika* 1997 (3): 65.

6. I. V. Ivkin, *Gosudarstvennaia vlast' SSSR* (Moscow, 1999), 194.

7. My description of the approach through the Kremlin to Stalin's office matches the one provided by Milovan Djilas in *Conversations with Stalin* ((New York, 1962), 58–59. For the dates of Aleksandrov's meetings with Stalin, see "Posetiteli kremlovskogo kabineta I.V. Stalina—zhurnaly (tetrady) zapisi lits, priniatykh pervym gensekom, 1924–1953 g.g.," *Istoricheskii arkhiv* 1996 (4): 114–130 and 1998 (4): 18–19.

8. *Istoriia filosofii* (Moscow, 1941–1943).

9. RGASPI, f. 17, op. 125, d. 254, ll. 1–30.

10. Ibid., ll. 6–8.

11. RGASPI, f. 88, op. 1, d. 970, ll. 12–14; f. 17, op. 3, d. 1050, ll. 145–152.

12. RGASPI, f. 88, op. 1, d. 970, ll. 12–14.

13. RGASPI, f. 17, op. 125, d. 221, ll. 1–2.

14. RGASPI, f. 17, op. 125, d. 454, ll. 80–81, and V. D. Esakov, "Toward a History of the 1947 Philosophy Discussion," in *Russian Studies in Philosophy* 1994 (spring): 10–11.

15. *Sovetskaia kniga* 1946 (5): 65–69, and *Vestnik Akademii nauk* 1946 (10): 121–124.

16. RGASPI, f. 17, op. 125, d. 478, ll. 24–27.

17. Werner G. Hahn, *Postwar Soviet Politics: The Fall of Zhdanov and the Defeat of Moderation, 1945–1953* (Ithaca, 1982), 19–66.

18. Pospelov took notes during the meeting, and from them we can piece together what Stalin said. RGASPI, f. 629, op. 1, d. 54, ll. 23–30. The names of the participants and the time of the meeting listed in Pospelov's notes correspond precisely to those found in Stalin's archive. See "Posetiteli" in footnote 7.

19. Ibid.

20. David Joravsky, *Soviet Marxism and Natural Science, 1917–1921* (New York, 1961), 170–205.

21. RGASPI, f. 629, op. 1, d. 54, ll. 23–30.

22. Ibid.

23. RGASPI, f. 17, op. 117, d. 681, l. 147; f. 17, op. 125, d. 477, ll. 1–2; Esakov, "Toward a History," 15.

24. RGASPI, f. 17, op. 125, d. 478, ll. 6–10, and Arkhiv Rossiiskoi Akademii Nauk (ARAN), f. 1922, op. 1, d. 233, l. 140.

25. RGASPI, f. 17, op. 125, d. 478, l. 12.

26. ARAN, f. 1922, op. 1, d. 234, l. 33.

27. ARAN, f. 1922, op. 1, d. 233, ll. 1–24.

28. Ibid., ll. 25–36.

29. RGASPI, f. 17, op. 125, d. 477, ll. 4–12.

30. ARAN, f. 1922, op. 1, d. 233, ll. 47–81.

31. ARAN, f. 1922, op. 1, d. 235, ll. 50–86.

32. RGASPI, f. 17, op. 125, d. 477, ll. 13–17 and 118–131.

33. Ibid., ll. 18–39.

34. ARAN, f. 1922, op. 1, d. 235, ll. 100–101.

35. ARAN, f. 1922, op 1, d. 234, l. 37.

36. Ibid., ll. 143–150. The author of this speech, Z. A. Kamenskii, was not allowed to address the audience. His speech was submitted in written form.

37. Ibid., l. 73.

38. ARAN, f. 1922, op. 1, d. 235, ll. 112–127.

39. Ibid., ll. 128–134.40. RGASPI, f. 17, op. 125, d. 478, ll. 6–23. The reports were also signed by Vasetskii, the director of the Institute of Philosophy.

41. M. G. Iaroshevskii in *Repressirovannaia nauka: Vypusk 1* (Leningrad, 1991).

42. RGASPI, f. 17, op. 125, d. 478, ll. 28–51, and Esakov, "Toward a History," 17–18.

43. RGASPI, f. 17, op. 125, d. 478, ll. 52–79.

44. RGASPI, f. 17, op. 3 d. 1064, l. 45; and for a draft of the decree, see f. 17, op. 163, d. 1065, l. 234.

45. RGASPI, f. 17, op. 125, d. 479, ll. 1–41.

46. *Voprosy filosofii* 1947 (1): 6–7. The first issue of *Voprosy filosofii* (hereafter VF) published the minutes from the discussion.

47. Ibid., 13, 109.

48. Esakov, "Toward a History," 20–22.

49. VF 1947 (1): 120–130.

50. See speeches by Leonov, ibid., 151–158; Gak, 19–26; and Nikolaev, 206–209.

51. See speches by Shariia, ibid., 163–171; Kovchegov, 245–248; Serebriakov, 101; Kedrov, 38–53; Kammari, 13–19; and Buinitskii, 232–237.

52. Ibid., 261–268.

53. Ibid., 63–64.

54. See RGASPI, f. 17, op. 125, d. 492, ll. 35–37. The speaker was Kh. G. Adzheminan. His speech was not included in the published minutes. For details, see Esakov, "Toward a History," 35.

55. RGASPI, f. 17, op. 125, d. 404, ll. 79–85.

56. ARAN, f. 1922, op. 1, d. 234, l. 173.

57. RGASPI, f. 17, op. 117, d. 870, ll. 69–83.

58. VF 1947 (1): 52–53.

59. Ibid., 64–65.

60. Ibid., 267–268. B. M. Kedrov, "Kak sozdavalsia nash zhurnal," VF 1988 (4): 95. Kedrov's memoirs correspond well with information available in the archives. He is wrong, however, in asserting that the title of the journal was first proposed at a meeting with Zhdanov in the summer of 1947. Documents from the Academy of Sciences and the Party indicate that "Voprosy filosofii" was the working title for the journal from the beginning.

61. VF 1947 (1): 41–42.

62. Ibid., 70–71.

63. Ibid., 82–83 and 134.

64. Ibid., 220.

65. Ibid., 192–196.

66. Ibid., 441.

67. Ibid., 271.

68. Ibid.

69. Ibid., 103.

70. Ibid., 268.

71. Ibid., 237–240.

72. Ibid.: see speeches by Zakhidov, 91–97; Chaloian, 142–147; Makolevskii, 209–212; and Sarabianov, 130–136.

73. Ibid.: see speeches by Emdin, 6–13; Zhukov, 80–83; Mitin, 120–130; and Iovchuk, 212–221.

74. Ibid., 109–113. The quote is from Smirnova's speech, 109. Nearly her whole talk was dedicated to the theme of integrating Russian thought into the history of Europe.

75. Ibid., 38–53.

76. *Bolshevik* 1945 (21): 45.

77. VF 1947 (1): Svetlov, 54–66, and Iovchuk, 212–221.

78. For the strongest rebuttal to Iovchuk's emphasis on the role of Russian tradition in Leninism and Soviet Marxism, see Selektor's planned speech for the meeting. Ibid., 427–436.

79. Ibid., 260.

80. Ibid., 288–299.

81. Quoted in Batygin, "Delo Akademika," *Chelovek* 1993 (1): 145.

82. Kedrov, "Kak sozdavalsia," 95.

83. RGASPI, f. 17, op. 125, d. 492, ll. 79–86.

84. Kedrov, "Kak sozdavalsia," 102.

85. RGASPI, f. 17, op. 132, d. 160, ll. 93–98.

86. Ibid.

87. I. N. Zhukov, "Bor'ba za vlast' v rukovodstve SSSR v 1945–1952 godakh," *Voprosy istorii* 1995 (1): 23–39.

CHAPTER 3
"THE FUTURE BELONGS TO MICHURIN"

1. Lysenkoism is perhaps the most studied aspect of Soviet science. Lysenko is mentioned in every major textbook and has been the subject of numerous articles and books. Early interpretations, which pitted "dialectical materialism" against "science" include P. S. Hudson and R. S. Richens, *The New Genetics in the Soviet Union* (Cambridge, 1946); C. Zirkle, *Death of a Science in Russia* (Philadelphia, 1949); and J. Huxley, *Soviet Genetics and World Science* (London, 1949). David Joravsky's *Lysenko Affair* (Cambridge, 1970) turned away from this approach and looked instead to the political, social, and cultural context to explain Lysenko's rise and the parallel fall of genetics. A series of other scholars have attempted to show that Lysenkoism was in fact exemplary of the working of "Stalinist science" in general. See Valerii Soifer, *Vlast' i nauka: istoriia razgroma genetiki v SSSR* (Tenafly, New Jersey, 1989), and Valery Soyfer, *Lysenko and the Tragedy of Soviet Science* (New Brunswick, 1994). For a sophisticated exploration of the political tactics shared by Lysenko and the geneticists, see Nikolai Krementsov, *Stalinist Science* (Princeton, 1997). The most impressive archival findings have been published by V. Esakov, S. Ivanova, and E. Levina, "Iz istorii bor'by s lysenkovshchinoi," *Izvestiia TsK KPSS* 1991 (4): 125–141, 1991 (6): 157–173, and 1991 (7): 109–

121, and by Kiril Rossianov, "Editing Nature: Joseph Stalin and the 'New' Soviet Biology," *Isis* 1993 (84): 728–745. For the first Russian account of Lysenkoism, see Zhores Medvedev's *Rise and Fall of T. D. Lysenko* (New York, 1969), which was written in the Soviet Union but, because Medvedev was critical of Lysenko and by implication the system, was initially published in the West. Finally, for a look at how the Lysenko affair was understood by Western European Marxists, see Dominique Lecourt, *Proletarian Science? The Case of Lysenko* (London, 1977).

2. Vladislav Zubok and Constantine Pleshakov, *Inside the Kremlin's Cold War: From Stalin to Khrushchev* (Cambridge, 1996), 110–137.

3. Joravsky, *Lysenko Affair*, 82–85; Soyfer, *Lysenko and the Tragedy*, 12–42.

4. Joravsky, *Lysenko Affair*, 40–54.

5. Soyfer, *Lysenko and the Tragedy*, 60–62.

6. Krementsov, *Stalinist Science*, 70–71.

7. Ibid., 74–80.

8. RGASPI, f. 17, op. 125, d. 547, ll. 1–5.

9. See A. R. Zhebrak, "Soviet Biology," *Science* 1945 (102): 357–358, and N. P. Dubinin, "Work of Soviet Biologists: Theoretical Genetics," *Science* 1945 (105): 109–112. In February 1945 Zhebrak wrote to Malenkov to ask permission to send his article to *Science* as a response to a recent article of the same title written by two American biologists. In the letter Zhebrak accused Lysenko of misleading people about the validity of genetics research.

10. *Politbiuro TsK VKP (b) i Sovet Ministrov SSSR, 1945–1953* (Moscow, 2002), 203–205.

11. Werner G. Hahn, *Postwar Soviet Politics: The Fall of Zhdanov and the Defeat of Moderation, 1945–1953* (Ithaca, 1982), 41–43. Yoram Gorlizki and Oleg Khlevniuk, *Cold Peace: Stalin and the Soviet Ruling Circle, 1945–1953* (Oxford, 2004), 26–31.

12. Amy Knight, *Beria: Stalin's First Lieutenant* (Princeton, 1993), 143.

13. RGASPI, f. 17, op. 125, d. 547, ll. 126–132.

14. Krementsov, *Stalinist Science*, 148.

15. *Politbiuro*, 269.

16. Soyfer, *Lysenko and the Tragedy*, 168.

17. "Obmen pis'mami mezhdu T. D. Lysenko i I. V. Stalinym v oktiabre 1947 g.," *Voprosy istorii estestvoznaniia i tekhniki* 1998 (2): 157–161.

18. Ibid., 161–163.

19. Ibid., 164. Stalin's name appears in all capital letters in the original letter.

20. Ibid.

21. Ibid., 163–164.

22. Ibid.

23. Ibid., 165.

24. Soyfer, *Lysenko and the Tragedy*, 167.

25. RGASPI, f. 17, op. 125, d. 547, ll. 4–5.

26. Articles supporting Lysenko appeared throughout the fall of 1947 in the *Literary Gazette*. See especially the interview with Lysenko in which he attacked Zhukovskii for believing in intraspecies competition. "Pochemu burzhuaznaia nauka vosstaet protiv rabot sovetskikh uchenykh," *Literaturnaia gazeta*, October 18, 1947.

27. RGASPI, f. 17, op. 125, d. 619, ll. 2–14.

28. Ibid., ll. 43–48.

29. See I. I. Shmal'gauzen, "Predstavleniia o tselom v sovremennoi biologii," *Voprosy filosofii* 1947 (2): 177–183. The journal refrained from clearly taking sides in the discussion and also published a brief notice of a meeting of Lysenko and members of the editorial board of the journal. It was agreed by all that the philosophical questions of biology deserved serious attention from biologists and professional philosophers alike. See I. I. Novinskii, "Besseda s akad. T. D. Lysenko v redaktsii zhurnala 'Voprosy filosofii,' " in the same number of the journal, 384.

30. RGASPI, f. 17, op. 125, d. 619, ll. 17–20.

31. Ibid., ll. 15–16.

32. Ibid., ll. 21–22.

33. RGASPI, f. 17, op. 138, d. 33, ll. 2–50. Tsitsin may have sensed in the spring of 1948 that Lysenko was on the defensive. If so, he clearly guessed wrong. The letter resurfaced in the fall of 1948 when Lysenko reported to Malenkov on his progress with branched wheat. Malenkov asked the head of the Agricultural Section of the Central Committee to discuss the situation with him, but there is no trace of what they said or whether Malenkov was skeptical about Lysenko's reports. For more on Tsitsin's letter, see Douglas Weiner, *A Little Corner of Freedom: Russian Nature Protection from Stalin to Gorbachev* (Berkeley, 1999), 73–75.

34. RGASPI, f. 17, op. 125, d. 619, ll. 62–67.

35. Krementsov, *Stalinist Science*, 164.

36. A copy of Yuri Zhdanov's speech is contained in RGASPI, f. 17, op. 125, d. 620, ll. 2–45.

37. Tsentral'nyi Gosudarstvennyi arkhiv obshchestvennykh dvishenni goroda Moskvy (TsAODM), f. 2501, op. 1, d. 5, ll. 2–4.

38. This letter was published in Soyfer, *Lysenko and the Tragedy*, 173–175. The translation from Soyfer's book is used here.

39. Krementsov, *Stalinist Science*, 165.

40. "Posetiteli kremlevskogo kabineta I. V. Stalina: 1947–1949," *Istoricheskii arkhiv*, 1996 (5–6): 35.

41. Quoted in Krementsov, *Stalinist Science*, 166. The quotation comes from Yuri Zhdanov's recollection of the meeting. See Yuri Zhdanov, "Vo mgle protivorechii," *Voprosy filosofii* 1993 (7): 86–87.

42. Shepilov's memoirs are quoted in Soyfer, *Lysenko and the Tragedy*, 180.

43. From V. A. Malyshev's diary, published in *Istochnik* 1997 (5): 135.

44. See Esakov et al., "Iz istorii," *Izvestiia TsK KPSS* 1991 (7): 112, and Krementsov, *Stalinist Science*, 166–167. The original notes are in Zhdanov's archives, RGASPI, f. 77, op. 3a, d. 180, l. 21. It is not clear exactly when Zhdanov took these notes, because they are not dated.

45. The memo runs twenty-seven pages and was subsequently distributed by Malenkov to other Politburo members. RGASPI, f. 17, op. 125, d. 620, ll. 105–132.

46. I. V. Stalin, *Sochineniia*, vol. 1, p. 303.

47. RGASPI, f. 558, op. 11, d. 158, l. 72. I would like to thank Olga Shevchenko for helping me decipher Stalin's handwriting in this document.

48. Both the initial draft with A. A. Zhdanov's editorial remarks and the revised version are in Zhdanov's archive. RGASPI, f. 77, op. 1, d. 991, ll. 1–25 and 26–46.

49. As is discussed later in the chapter, the letter was eventually published in *Pravda*, August 7, 1948. I use the translation from Soyfer, *Lysenko and the Tragedy*, 190–191.

50. Ibid.

51. Hahn, *Postwar Soviet Politics*, 109–113.

52. RGASPI, f. 17, op. 163, d. 1513, l. 52, and f. 17, op. 3, d. 1071, l. 24.

53. RGASPI f. 558, op. 11, d. 158, ll. 23–44. This suggests that Yuri Zhdanov's speech was central to the timing of the Agricultural Academy meeting and raises questions about Krementsov's argument that "by far the most important factor in Stalin's decision to intervene on Lysenko's behalf in July 1948 was the escalating Cold War." He cites the increasing tensions over Yugoslavia and Germany in the summer of 1948. As Krementsov sees it, Stalin was using the Lysenko affair "as a pretext to announce a new party line in domestic and foreign policy." I see the events as incidentally, not causally, related. A speech given by Lysenko seems a particularly clumsy way to announce a new domestic and foreign policy. In my view, the timing can be explained by the internal momentum of the discussion as it unfolded in the Party, press, and academy from 1947 to 1948. See Krementsov, *Stalinist Science*, 159.

54. RGASPI f. 17, op. 3, d. 1071, l. 31. A copy of the earlier draft is in f. 17, op. 163, d. 1513, ll. 106–110.

55. The letter was published in Esakov, "Iz istorii," *Izvestiia TsK KPSS* 1991 (7): 119–120.

56. "Posetiteli," 41.

57. This point is made most clearly by Kirill O. Rossianov. Stalin's version of Lysenko's draft was discovered in RGASPI in the early 1990s. The material was first published by V. Esakov, S. Ivanova, and E. Levina, "Iz istorii bor'by," and the most detailed analysis in English is Rossianov's "Editing Nature."

58. Rossianov, "Editing Nature," 737–738.

59. Ibid., 733–736.

60. Ibid., 744.

61. For a complete list of participants, see Rossiiskii gosudarstvennyi arkhiv ekonomiki (RGAE), f. 8390, op. 1, d. 2241, ll. 1–31.

62. *Pravda*, July 28, 1948.

63. RGAE, f. 8390, op. 1, d. 2241, ll. 1–31.

64. *The Situation in the Biological Sciences* (New York, 1949), 71 and 105, for example. References are made to this 1949 English translation of the stenographic record of the meeting. On occasion I have adjusted the translations for clarity.

65. Ibid., 163, 506, and 229.

66. Ibid., 253 and 276.

67. Krementsov, *Stalinist Science*, 130–131.

68. *The Situation*, 64, 244, 274.

69. Ibid., 574 and 502.

70. Ibid., 309 and 383.

71. Ibid., 89, 263, 264.

72. Ibid., 63, 194, 295, 331, 417, 420, 422, 424, 522.

73. Ibid., 63, 120, 181, 191, 227, 297, 317, 329, 368, 405.

74. Soyfer, *Lysenko and the Tragedy*, 185.

75. *The Situation*, 154–160. For a translation of Rapoport's effort to defend himself at the meeting, see Soyfer, *Lysenko and the Tragedy*, 186.

76. *The Situation*, 441–456.

77. Ibid., 426–440.

78. Ibid., 467–476.

79. Ibid., 456–467.

80. Ibid., 489–495.

81. Compare Nemchinov's speech in RGAE, f. 8390, op. 1, d. 2246, ll. 184–197, with the published record in *The Situation*, 555–562 and *O polozhenii v biologicheskoi nauke. Stenograficheskii otchet sessii VASKhNILa 31 iiulia–7 avgusta 1948 g.* (Moscow, 1948), 469–476.

82. Ibid.

83. RGAE, f. 8390, op. 1, d. 2241, ll. 68; *O polozhenii*, 281–302.

84. *The Situation*, 360.

85. RGASPI, f. 17, op. 132, d. 40, l. 4.

86. Ibid., ll. 1–2.

87. *Pravda*, August, 7, 1948.

88. The original note is full of abbreviations and Stalin's editorial marks. See Rossianov, "Editing Nature," 743–744.

89. *The Situation*, 618–620.

90. Ibid., 620–622.

91. Ibid., 623–624.

92. The history of the impact of the the Agricultural Academy meeting on Soviet biology has been told many times, so I will keep this account brief. For the best and most detailed analysis of the aftermath of the meeting in Soviet biology, see Joravsky, *Lysenko Affair*, chapters 5 and 6, and Krementsov, *Stalinist Science*, chapters 7 and 8.

93. RGASPI, f. 17, op. 118, d. 119, ll. 1, 33–65.

94. RGASPI, f. 17, op. 118, d. 120, ll. 1–56.

95. "Vyshe znamia peredovoi, michurinskoi biologicheskoi nauki!" *Pravda*, August 12, 1948.

96. RGASPI, f. 17, op. 118, d. 129, ll. 82–88.

97. Krementsov, *Stalinist Science*, 249–253.

98. Yuri Zhdanov later remarked that it is important to remember that genetics was not a well-developed field in 1948. In hindsight it became clear to him that neo-Lamarckian assumptions are false, but at the time he believed that Lysenko indeed could have been right. Stalin evidently was convinced. Personal interview with Yuri Zhdanov, July 24, 1995, in Rostov, Russia.

CHAPTER 4
"WE CAN ALWAYS SHOOT THEM LATER"

1. ARAN, f. 2, op. 1–48, d. 188, ll.1–4.

2. David Holloway, *Stalin and the Bomb* (New Haven, 1994), 138–140.

3. For more on Soviet physicists in the 1920s and 1930s, see Paul Josephson, *Physics and Politics in Revolutionary Russia* (Berkeley, 1991), and Karl Hall, "Purely Practical Revolutionaries: A History of Stalinist Theoretical Physics," Ph.D. dissertation, Harvard University, 1999.

4. G. E. Gorelik, "Fizika universitetskaia i akademicheskaia," *Voprosy istorii estestvoznaniia i tekhniki*, 1991 (2): 31–46

5. Ibid.

6. RGASPI, f. 17, op. 125, d. 363, ll. 57–59.

7. RGASPI, f. 17, op. 125, d. 445, ll. 175, 140–141. Kurchatov's letter was also signed by the head of the chemical industry and future Politburo member M. G. Pervukhin and by the administrative head of the atomic project for the Council of Ministers, B. L. Vannikov.

8. Gorelik, "Fizika," 31–46.

9. RGASPI, f. 17, op. 125, d. 363, l. 73–74.

10. RGASPI, f. 17, op. 125, d. 361, ll. 19–36.

11. Gorelik, "Fizika," 31–46. The quotation is from S. T. Konobeevskii, who was dean of physics at MGU from 1946 until the spring of 1947.

12. *Voprosy filosofii* 1947 (1): 271.

13. RGASPI, f. 17, op. 125, d. 492, ll. 13–17.

14. For a thorough introduction to the relationship between Soviet Marxist philosophy and the natural sciences, see David Joravsky, *Soviet Marxism and Natural Science, 1917–1932* (New York, 1961), and Loren Graham, *Science and Philosophy in the Soviet Union* (New York, 1972), chapters 1 and 2. For a detailed discussion of various Soviet positions on quantum mechanics, see Graham's "Quantum Mechanics and Dialectical Materialism," *Slavic Review* 1966 (3): 381–420.

15. Werner Heisenberg, *Physics and Philosophy: The Revolution in Modern Physics* (New York, 1958), 59. For an introduction to the debates in the West, see Daniel Kevles, *The Physicists* (Cambridge, 1995), and the chapter "The Earliest Missionaries of the Copenhagen Spirit" in John Heilbron, *Science in Reflection* (New York, 1988). For a discussion of the nexus between philosophy and physics at the time quantum theory was being formulated, see Paul Forman, "Weimar Culture, Causality and Quantum Theory, 1918–1927: Adaptation by German Physicists to a Hostile Intellectual Environment," *Historical Studies in the Physical and Biological Sciences* 1971 (3).

16. Graham, "Quantum Mechanics," and Alexander Vucinich, "Soviet Physicists and Philosophers in the 1930s: Dynamics of a Conflict," *Isis* 1980 (257): 236–250.

17. V. I. Lenin, *Materializm i empiriokrititsizm* (Moscow, 1946), 243. See also Joravsky, *Soviet Marxism*, chapters 1 and 2.

18. Vucinich, "Soviet Physicsts," 250.

19. Ibid., 239–241.

20. RGASPI, f. 17, op. 125, d. 363, ll. 64–74.

21. *Voprosy filosofii* 1947 (1): 440–444.

22. RGASPI, f. 77, op. 1, d. 983, ll. 180–184.

23. *Voprosy filosofii* 1947 (2): 138–139.

24. Alexei Kojevnikov, "President of Stalin's Academy: The Mask and Responsibility of Sergei Vavilov," *Isis* 1996 (87): 41–42.

25. *Voprosy filosofii* 1947 (2): 140–176.

26. A. A. Maksimov, "Ob odnom filosofskoi kentavre," *Literaturnaia gazeta*, April 10, 1948.

27. TsAODM, f. 2501, op. 1, d. 7, ll. 88–91. Aleksandrov also suggested holding a conference on philosophical questions of modern chemistry.

28. P. L. Kapitsa, "Ob organizatsii nauki," *Pod znamenem marksizma* 1943 (7–8): 90–101.

29. RGASPI, f. 17, op. 125, d. 221, l. 89.

30. RGASPI, f. 17, op. 3, d. 1066. The Politburo decision was finalized on June 16, 1947.

31. RGASPI, f. 77, op. 1, d. 989, l. 3.

32. RGASPI, f. 17, op. 118, d. 244, ll. 237–240.

33. Gosudarstvennyi arkhiv Rossiiskoi Federatsii (GARF), f. 9396, op. 1, d. 123, ll. 161–163. See also RGASPI, f. 17, op. 118, d. 244, ll. 237–240.

34. RGASPI, f. 17, op. 118, d. 262, ll. 58–60.

35. GARF, f. 9396, op. 1, d. 244, ll. 1–101.

36. GARF, f. 9396, op. 1, d. 245, ll. 1–60.

37. Ibid., ll. 59–60.

38. GARF, f. 9396, op. 1, d. 246, l. 251.

39. GARF, f. 9396, op. 1, d. 265, ll. 19–48; and d. 246, ll. 190–262.

40. GARF, f. 9396, op. 1, d. 246, ll. 192–246. Fock, Kedrov, Vul, and Leontovich were among those who responded to Maksimov.

41. Ibid., l. 248.

42. Ibid., ll. 252–253.

43. Ibid., ll. 254–255.

44. Ibid., l. 156. The initial responses to Maksimov's speech are given on pages 192–246 of the folder. Kaftanov's response is recorded on pages 247–259.

45. *Pravda* January, 29, 1949; See also *Kul'tura i zhizn'*, January 30, 1949, Feburary 11, 1949, Feburary 20, 1949, March 10, 1949, and March 22, 1949.

46. GARF, f. 9396, op. 1, d. 249, ll. 1–57.

47. GARF, f. 9396, op. 1, d. 250.

48. GARF, f. 9396, op. 1, d. 244, ll. 17–26.

49. ARAN, f. 2, op. 1–49, d. 141, ll. 1–2; RGASPI, f. 17, op. 118, d. 302, ll. 146–148.

50. RGASPI, f. 17, op. 132, d. 211, ll. 1–32.

51. RGASPI, f. 17, op. 118, d. 360, ll. 169–170.

52. RGASPI, f. 17, op. 132, d. 211, l. 3.

53. The various historical interpretations are discussed in V. P. Vizgin, "Iadernyi shchit v 'tridtsatiletnei voine' fizikov s nevezhestvennoi kritikoi sovremennykh fizicheskikh teorii," *Uspekhi fizicheskikh nauk* 1999 (12): 1371–1388. For a debate on the matter, see G. E. Gorelik and A. B. Kojevnikov, "Chto spaslo sovetskuiu fiziku ot lysenkovaniia? Dialog," *Priroda* 1999, 95–104. In this exchange Gorelik defends the traditional, memoir-based account in which Kurchatov saved physics from a Lysenko-like discussion. Kojevnikov, relying on archival documents, points to bureaucratic maneuvers in the Central Committee and Academy of Sciences.

54. From Holloway, *Stalin and the Bomb*, 211.

55. ARAN, f. 1922, op. 1, d. 439.

56. V. D. Esakov, "Mify i zhizn'," *Nauka i zhizn'* 1991 (11).

57. Personal interview with Yuri Zhdanov, July 24, 1995, in Rostov-on-Don, Russia.

58. RGASPI, f. 17, op. 132, d. 354, ll. 149–150.

59. Ibid., ll. 114–118. See also I. V. Kuznetsov's review in *Bol'shevik* 1950 (6).

60. See, for example, RGASPI, f. 17, op. 133, d. 230.

61. ARAN, f. 1922, op. 1, d. 439, ll. 1–125, d. 440, ll. 1–10.

62. A. A. Maksimov et al., *Filosofskie voprosy sovremennoi fiziki* (Moscow, 1952), 4.

63. Gorelik, "Fizika," 37.

64. V. A. Fock, "O tak nazyvaemykh ansambliakh v kvantovoi mekhanike," *Voprosy filosofii* 1952 (4): 170–174.

65. Loren Graham, "Quantum Mechanics," 399.

66. A. A. Maksimov, "Protiv reaktsionnogo Einshteinianstva v fizike," *Krasnyi flot*, June 13, 1952.

67. RGASPI, f. 17, op. 133, d. 256, ll. 1, 8–9.

68. "Beriia i teoriia otnositel'nosti," *Istoricheskii arkhiv* 1994 (3): 217.

69. Ibid., 217–218.

70. Ibid., 219.

71. Ibid., 220–221.

72. Ibid., 221–222.

73. RGASPI, f. 17, op. 133, d. 290, ll. 98–99.

74. For a fuller discussion of this institutional battle, see A. V. Andreev, "Sotsial'-naia istoriia NIIF MGU (1922–1994)," Ph.D. dissertation, Institute for the History of Science and Technology of the Russian Academy of Sciences (Moscow, 1994). The quotes are from Vizgin, "Iadernyi," 1383–1384.

75. Vizgin, "Iadernyi," 1383–1384.

76. Ibid.

77. Ibid.

78. Holloway, *Stalin and the Bomb*, 212–213.

79. Quoted in G. Kostyrchenko, *V plenu krasnogo faraona* (Moscow, 1994), 285.

80. RGASPI, f. 17, op. 132, dd. 191–195.

81. RGASPI, f. 17 op. 119, d. 138, ll. 31, 33.

82. Ibid., ll. 28–30, 34.

83. RGASPI, f. 17, op. 132, d. 211, l. 123.

84. Quoted in Andrei Sakharov, *Memoirs* (New York, 1990), 209. Anna Alekseevna Kapitsa's quotation comes from an interview with her in the documentary film *Citizen Kurchatov: Stalin's Bomb Maker*.

85. Kojevnikov, "President," 47–48.

86. See Loren Graham, "Do Mathematical Equations Display Social Attributes?" *Mathematical Intelligencer* 2000 (22): 31–36. Here Graham convincingly argues that Fock's dialectical materialism influenced his mathematical equations.

87. These letters are published in G. E. Gorelik's article, "V. A. Fock: Filosofiia tiagoteniia i tiashest' filosofii," *Priroda* 1993 (3): 87–88.

88. Ibid., 88.

CHAPTER 5
"A BATTLE OF OPINIONS"

1. *Pravda*, June 27, 1950, 3–4.

2. Katerina Clark, *Petersburg: Crucible of Cultural Revolution* (Cambridge, 1995): 216; See also Yuri Slezkine, "N. Ia. Marr and the National Origins of Soviet Ethnogenetics," *Slavic Review* 1996 (4): 831.

3. A number of outstanding articles and monographs deal with N. Ia. Marr's life and scholarship. See Lawrence L. Thomas's *Linguistic Theories of N. Ja. Marr* (Berkeley, 1957) and, for a briefer account, his "Some Notes on the Marr School," *American and Slavic and East European Review* 1957 (3): 324–348. For another early description of Marr and Soviet linguistics, see J. Ellis and R. W. Davies, "The Crisis in Soviet Linguistics," *Soviet Studies* 1951 (3): 209–264. I have also greatly benefited from more general overviews of Marr's work, including Slezkine, "N. Ia. Marr," 826–862; chapter 4 in Vera Tolz, *Russian Academicians and the Revolution: Combining Professionalism and Politics* (London, 1997); pages 212–223 in Clark, *Petersburg*; and especially V. M. Alpatov, *Istoriia Odnogo Mifa: Marr i Marrizm* (Moscow, 1991). For this chapter I have borrowed from each of these sources to summarize what is obviously a complicated story of Marr, Marrism, and their relation to Soviet linguistics in the 1920s and 1930s.

4. Tolz, *Russian Academicians*, 89.

5. Quoted in Slezkine, "N. Ia. Marr," 843.

6. Ibid., 856.

7. Izvestiia Akademii Nauk SSSR: otdelenie literatury i iazyka (*IANOLIa*) 1946 (4 and 5); *IANOLIa* 1947 (4): 381–394.

8. Ibid.

9. B. Agapov and K. Zelinskii, "Net eto ne russkii iazyk," *Literaturnaia gazeta*, December 17, 1947; G. Serdiuchenko, "Za samokritiku v sovetskom iazykovedenii," *Literaturnaia gazeta*, December 17, 1947; and Alpatov, *Istoriia*, 144–145.

10. I. I. Meshchaninov, "O polozhenii v lingvisticheskoi nauke," *IANOLIa* 1948 (6): 473–474.

11. Ibid., 473–485.

12. F. P. Filin, "O dvukh napravleniiakh v iazykovedenii," *IANOLIa* 1948 (6): 486.

13. Ibid., 486–496.

14. "Rezoliutsiia," *IANOLIa* 1948 (6): 497–499.

15. "Za partiinost' v nauke o iazyke i literature," *Literaturnaia gazeta*, October 13, 1948, and "Protiv idealizma i nizkopoklonstva v iazykoznanii," *Literaturnaia gazeta*, November 17, 1948.

16. *IANOLIa* 1949 (4): 393.

17. TsAODM, f. 2505, op. 1, d. 1, l. 218.

18. N. Bernikov and I. Braginov, "Za peredovoe sovetskoe iazykoznanie," *Kul'tura i zhizn'*, May 11, 1949, and G. Serdiuchenko, "Ob odnoi vrednoi teorii v iazykoznanii," *Kul'tura i zhizn'*, June 30, 1949.

19. RGASPI, f. 17, op. 132, d. 164, ll. 16–33.

20. Ibid., ll. 34–39.

21. Ibid., ll. 42–43.

22. ARAN, f. 2, op. 1–50, d. 12, ll. 1–2.

23. RGASPI, f. 17, op. 132, d. 336, ll. 11–13.

24. Arn. Chikobava, "Kogda i kak eto bylo," *Ezhegodnik iberiisko-kavkazskogo iazykoznaniia* 1985 (12): 9–14.

25. Stalin's copy of the letter is in his *fond* at RGASPI, f. 558, op. 11, d. 1250, ll. 2–14.

26. Ibid. Nui Jun, "The Origins of the Sino-Soviet Alliance," in Odd Arne Westad, ed., *Brothers in Arms: The Rise and Fall of the Sino-Soviet Alliance, 1945–1963* (Stanford,

1998), and Evgueni Bajanov, "Assessing the Politics of the Korean War, 1949–1951," *Cold War International History Project Bulletin* 1995/1996 (Winter, 6–7): 87.

27. V. S. Ilizarov, "Dokumental'nye ocherki. Pochetsnyi akademik I. V. Stalin protiv akademika N. Ia. Marra. K istorii diskussii po voprosam iazykoznaniia v 1950 g.," *Novaia i noveishaia istoriia* 2003 (7): 112–113.

28. Chikobava, "Kogda," 9–14. Roy Medvedev relays the conversation slightly differently, without citing his sources. According to him, Stalin asked: "And who are these people?" Arutiunov responded: "They are scientists, academicians. . . ." Stalin retorted: "And I thought they must be accountants since they were so easily removed from one place and placed in another. You rushed things, Comrade Arutiunov, you rushed things." See Roy Medvedev, "Stalin i iazykoznanie," *Vestnik Rossiiskoi Akademii Nauk* 1997 (67): 1037.

29. Chikobava, "Kogda," 9–14. Slezkine, "N. Ia. Marr," 857. Chikobava's account is neither supported nor explicitly contradicted by the record of visitors to Stalin's office published in "Posetiteli kremlevskogo kabineta I. V. Stalina—zhurnal (tetrady) zapisi lits, priniatykh pervym gensekom, 1924–1953 g.g.," *Istoricheskii arkhiv* 1997 (1): 11. Since Chikobava says he met with Stalin at his dacha, the Kremlin office records cannot corroborate that the meeting took place. However, it is clear that Charkviani met with Stalin on the night of April 12 at the Kremlin, thus confirming Chikobava's claim that his patron was in Moscow on official business at the time.

30. RGASPI, f. 558, op. 11, d. 1251, contains three drafts of the article with Stalin's corrections on listy 122–163, 62–121, and 3–61.

31. Ibid., ll. 4–6.

32. Ibid., ll. 2–3.

33. Ibid., ll. 1–3.

34. For this and subsequent articles in *Pravda* on linguistics I will cite both the issue of *Pravda* and the page in Ernest J. Simmons, ed., *The Soviet Linguistics Controversy* (New York, 1951), which contains excellent English translations by John V. Murra, Robert M. Hankin, and Fred Holling. I occasionally make minor alterations to their translations for clarity and readability. Simmons, *Linguistics*, 9; *Pravda* May 9, 1950, 3.

35. Simmons, *Linguistics*, 9–19; *Pravda* May 9, 1950, 3–5.

36. Ibid.

37. Medvedev, "Stalin," 1037.

38. RGASPI, f. 17, op. 132, d. 337, l. 164.

39. RGASPI, f. 17, op. 118, d. 875, ll. 57–64.

40. Ibid., ll. 53–56.

41. ARAN, f. 678, d. 2, op. 105, l. 74. Though proposed in May, the decision was not enacted until after the conclusion of the discussion.

42. *Pravda*, May 16, 1950, 3–4; Simmons, *Linguistics*, 20–28.

43. Ibid.

44. *Pravda*, May 23, 1950, 4; Simmons, *Linguistics*, 36–38.

45. *Pravda*, June 6, 1950, 3–4; Simmons, *Linguistics*, 48.

46. *Pravda*, May 30, 1950, 3–4; Simmons, *Linguistics*, 42–45.

47. Ibid. *Pravda*, June 6, 1950, 3–4; Simmons, *Linguistics*, 48. *Pravda*, May 23, 1950, 3–4; Simmons, *Linguistics*, 32–35. *Pravda*, June 13, 1950, 4; Simmons, *Linguistics*, 62–64.

48. *Pravda*, May 23, 1950, 3–4; Simmons, *Linguistics*, 32–35.

49. *Pravda*, May 23, 1950, 3; Simmons, *Linguistics*, 29–31.

50. *Pravda*, June 13, 1950, 4; Simmons, *Linguistics*, 62, 65.

51. *Pravda*, May 23, 1950, 4; Simmons, *Linguistics*, 37. *Pravda*, May 23, 1950, 3–4; Simmons, *Linguistics*, 35. *Pravda*, May 30, 1950, 4; Simmons, *Linguistics*, 46–48.

52. *Pravda*, June 6, 1950, 3–4; Simmons, *Linguistics*, 48–57.

53. *Pravda*, June 13, 1950, 3; Simmons, *Linguistics*, 61. *Pravda*, May 23, 1950, 4; Simmons, *Linguistics*, 36.

54. *Pravda*, May 30, 1950, 3; Simmons, *Linguistics*, 40.

55. Ilizarov, "Dokumental'nye ocherki," 130–135.

56. *Pravda*, June 20, 1950, 3–4; Simmons, *Linguistics*, 70–76.

57. Although the exact letters have not been found, files in Agitprop suggest that they may have been from N. S. Koshkina, identified as a student at LGU, and G. Krylov, identified as a student from MGU. These letters are listed as "responses" to Stalin's articles but were written on June 3 and June 10 respectively.

58. *Pravda*, June 20, 1950, 3–4; Simmons, *Linguistics*, 70–76.

59. Ibid.

60. Ibid.

61. Ibid.

62. Ibid.

63. *Pravda* June 27, 1950, 3–4; Simmons, *Linguistics*, 76–85. *Pravda*, July 4, 1950, 3–4; Simmons, *Linguistics*, 86–96.

64. RGASPI, f. 17, op. 132, d. 337, l. 2, identifies Krashennikova as a teacher at a Moscow pedagogical institute. *Pravda* received her letter on June 23, 1950.

65. *Pravda*, July 4, 1950, 3–4; Simmons, *Linguistics*, 86–96.

66. Ibid.

67. RGASPI, f. 17, op. 132, d. 337, l. 164.

68. Alpatov, *Istoriia*, 195.

69. RGASPI, f. 17, op. 3, d. 1083, ll. 25–26. RGASPI, f. 17, op. 118, d. 946, ll. 145–152; d. 955, ll. 2–8; d. 967, ll. 179–186; d. 976, l. 75.

70. RGASPI, f. 17, op. 118, d. 969, ll. 36–37; d. 970, ll. 232–250, and RGASPI, f. 17, op. 3, d. 1083, ll. 25–26.

71. RGASPI, f. 17, op. 118, d. 969, ll. 36–37; d. 970, ll. 232–250.

72. ARAN, f. 2, op. 1–50, d. 258, l. 26.

73. RGASPI, f. 17, op. 118, d. 942, ll. 105–109.

74. *Pravda*, August 2, 1950, 2; Simmons, *Linguistics*, 97.

75. Ibid.

76. Ilizarov, "Dokumental'nye ocherki," 111.

77. *Pravda*, August 2, 1950, 2; Simmons, *Linguistics*, 97.

78. RGASPI, f. 17, op. 132, d. 336, l. 141.

79. Ibid., l. 150.

80. RGASPI, f. 17, op. 132, d. 336, d. 337, d. 338, are full of letters to *Pravda*, the Central Committee, and Stalin about his articles on linguistics.

81. RGASPI, f. 17, op. 132, d. 338, ll. 245–247.

82. Ibid., ll. 247–249.

83. RGASPI, f. 17, op. 132, d. 337, ll. 185–187.

84. RGASPI, f. 558, op. 11, d. 116, ll. 8–45.

85. RGASPI, f. 17, op. 133, d. 6, ll. 30–32.

86. Ibid., ll. 33–43.

87. RGASPI, f. 17, op. 119, d. 483, l. 132; d. 526, ll. 68–71.

88. RGASPI, f. 17, op. 133, d. 223, l. 9; RGASPI, f. 17, op. 119, d. 582, ll. 41–44.

89. V. V. Vinogradov, ed., *Protiv vul'garizatsii i izvrashcheniia marksizma v iazykoznanii, sbornik statei, chast pervaia i vtoraia* (Moscow, 1951 and 1952).

90. RGASPI, f. 17, op. 133, d. 4, ll. 60–61.

CHAPTER 6
"ATTACK THE DETRACTORS WITH CERTAINTY OF TOTAL SUCCESS"

1. Vera Tolz, *Russian Academicians and the Revolution: Combining Professionalism and Politics* (London, 1997), 123–140.

2. Daniel Todes, "Pavlov and the Bolsheviks," *History and Philosophy of the Life Sciences* 1995 (1): 417–418. See also N. A. Grigor'ian, "Obshchestvenno-politicheskie vzgliady I. P. Pavlova," *Vestnik Akademii Nauk SSSR* 1991 (10): 74–87, and Tolz, *Russian Academicians,* 123–140. Pavlov's loyalty to the regime was first questioned in Soviet newspapers during glasnost. Grigor'ian used archival materials to challenge the official version of his political conversion in the mid-1930s. As Todes notes, once the old myths were questioned, there was a tendency to take a 180-degree turn and portray Pavlov as the "first dissident." For our purposes, it is important to note that, whatever Pavlov's actual position in the 1930s, after the Second World War the Party portrayed him as a loyal Soviet scientist.

3. David Joravsky, *Russian Psychology* (Oxford, 1989), and Loren Graham, *Science and Philosophy in the Soviet Union* (New York, 1972), and *Science, Philosophy, and Human Behavior in the Soviet Union* (New York, 1987), especially chapter 5, "Physiology and Psychology."

4. Robert C. Tucker, *The Soviet Political Mind: Studies in Stalinism and Post-Stalin Change* (New York, 1963), especially chapter 4, "Stalin and the Uses of Psychology."

5. Joravsky, *Russian Psychology,* 298.

6. Ibid., 286–307.

7. A. A. Zvorykin, ed., *Biograficheskii slovar' deiatelei estestvoznaniia i tekhniki* (Moscow, 1958), vol. 1, p. 66.

8. Ibid., vol. 2, p. 390.

9. V. L. Sviderskii, ed., *Akademik Leon Abgarovich Orbeli: nauchnoe nasledstvo* (Moscow, 1997); L. G. Leibson, *Akademik L. A. Orbeli: neopublikovannye glavy biografii* (Leningrad, 1990).

10. Joravsky, *Russian Psychology,* 399.

11. Joravsky, *Russian Psychology,* 388–389.

12. Elena Zubkova, *Russia after the War: Hopes, Illusions, and Disappointments, 1945–1957* (New York, 1998), 120. For more information on Orbeli's trip to England: Sviderskii, *Orbeli,* 117–130, 284–285.

13. Sviderskii, *Orbeli,* 24–25 and 32–33.

14. Nikolai Krementsov, *Stalinist Science* (Princeton, 1997), 266, and Joravsky, *Russian Psychology,* 296.

15. Krementsov, *Stalinist Science,* 264–266, and Joravsky, *Russian Psychology,* 296–297.

16. The Central Committee did not get around to approving the measure until December 1948. See RGASPI, f. 17, op. 118, d. 247.

17. Krementsov, *Stalinist Science*, 272.

18. Quoted in V. D. Esakov, ed., *Akademiia nauk v resheniiakh politbiuro TsK RKP (b)–VKP (b) 1922–1952* (Moscow, 2000), 372.

19. RGASPI, f. 17, op. 118, d. 151, ll. 6–15; f. 17, op. 132, d. 40, ll. 59–64.

20. RGASPI, f. 17, op. 132, d. 40, ll. 65–66, and RGASPI, f. 17, op. 118, d. 406, l. 59, and RGASPI, f. 17, op. 3, d. 1076, l. 32.

21. RGASPI, f. 17, op. 132, d. 169, ll. 43–60.

22. RGASPI, f. 17, op. 118, d. 717, ll. 11–12.

23. RGASPI, f. 17, op. 132, d. 177, l. 61.

24. RGASPI, f. 17, op. 132, d. 251, ll. 2–3.

25. RGASPI, f. 17, op. 132, d. 271, l. 71. K. Bykov, "Akademik Ivan Pavlov," *Kul'-tura i zhizn'*, February 20, 1949; "Akademik Ivan Pavlov," *Pravda*, Feburary 22, 1949.

26. RGASPI, f. 17, op. 118, d. 403, l. 38.

27. Esakov, *Akademiia*, 412–413. For details on the organization of the celebrations, see ARAN, f. 2, op. 1–49, d. 238–242.

28. RGASPI, f. 17, op. 132, d. 177, ll. 18–19.

29. Ibid.

30. "Velikii syn russkogo naroda," *Pravda*, September 27, 1949.

31. RGASPI, f. 83, op. 1, d. 9, l. 37; RGASPI, f. 17, op. 132, d. 177, l. 144.

32. RGASPI, f. 17, op. 132, d. 177, ll. 144–162.

33. Ibid., ll. 146–149.

34. Ibid., ll. 149–156.

35. Ibid., ll. 145–162.

36. Ibid., ll. 156–159.

37. Ibid., ll. 162.

38. RGASPI, f. 83, op. 1, d. 9, ll. 34–37.

39. RGASPI, f. 558, op. 11, d. 762, ll. 23–25; RGASPI, f. 83, op. 1, d. 9, ll. 34–37.

40. RGASPI, f. 17, op. 132, d. 347, ll. 9–10.

41. Ibid., ll. 1–4

42. Ibid.

43. RGASPI, f. 17, op. 118, d. 885, ll. 1–4; RGASPI, f. 17, op. 3, d. 1082, l. 13.

44. While it is clear that day-to-day decisions were being made by Kruzhkov and Zhdanov and at times they wrote directly to Malenkov and Stalin, Suslov was not out of the loop. See RGASPI, f. 17, op. 132, d. 347, ll. 5–7.

45. Krementsov, *Stalinist Science*, 273. RGASPI, f. 17, op. 132, d. 347, ll. 10–94.

46. RGASPI, f. 17, op. 118, d. 885, ll. 1–4; RGASPI, f. 17, op. 3, d. 1082, l. 13.

47. See ARAN, f. 2, op. 7, d. 76, ll. 1–47.

48. Ibid.

49. Ibid.

50. Ibid., l. 28.

51. Ibid., l. 37. Discussions of the numbers of people participating are interspersed throughout the minutes. See pages 12–47.

52. Ibid., ll. 32–35.

53. Ibid., ll. 13–14, 24.

54. Ibid., l. 16.

55. Ibid., ll. 18–20.

56. See Krementsov, *Stalinist Science* , chapter 9; Leibson, *Orbeli*, chapter 4.

57. ARAN, f. 2, op. 1–50, d. 254, ll. 11–15, 45–50; RGASPI, f. 17, op. 132, d. 347, ll. 111–117; *Nauchnaia sessiia posviashchennaia problemam fiziologicheskogo ucheniia Akademika I. P. Pavlova* (Moscow, 1951), 82. *Pravda*, June 29–July 9, 1950. For another analysis of this meeting see Joravsky, *Russian Psychology*, 406–414.

58. *Nauchnaia sessiia*, 5–8.

59. Ibid., 9–12.

60. Ibid., 13–43.

61. Ibid.

62. Ibid.

63. Ibid.

64. Ibid., 44–81.

65. Ibid., 95–102.

66. Ibid., 103–111.

67. Ibid., 116–125.

68. Ibid., 133–138.

69. Ibid., 160–164.

70. Ibid., 165.

71. Ibid.

72. Ibid., 166–173.

73. Ibid.

74. Ibid., 174–177.

75. Ibid., 177–184, 198–205, 392–396, 482–484. See also Leibson, *Orbeli*, 90–93.

76. *Nauchnaiia sessia*, 282–292.

77. Beritashvili sent Vavilov the text of a speech for inclusion in the record, but only in early July, after the session. Vavilov responded by telegram that the speech "was completely useless. . . . I ask you to inform me by telegraph whether you insist on publishing the speech." Beritashvili sent another version and asked that Vavilov determine whether it was worth publishing. It was eventually included in the record of the conference along with other speeches that had not actually been delivered. ARAN, f. 2, op. 1–50, d. 254, ll. 66, 121.

78. Gennadi Kostyrchenko, *Out of the Red Shadows: Anti-Semitism in Stalin's Russia* (New York, 1995), 131.

79. *Nauchnaia sessiia*, 405–408; 515.

80. Ibid., 298–303; 515.

81. Ibid., 106.

82. Ibid., 361–368.

83. Ibid., 501–504.

84. Ibid., 505–518.

85. Ibid., 516. The translation is provided in *Scientific Session of the Physiological Teachings of Academician I. P. Pavlov* (Moscow: 1951), 161: (Note: while the translation gets the point across, the word "Soviet" does not appear in the Russian version.)

86. *Nauchnaia sessiia*, 519–520.

87. RGASPI, f. 17, op. 133, d. 180, l. 30; f. 17, op. 119, d. 41, ll. 188–189; f. 17, op. 119, d. 42, ll. 20–25.

88. Ibid.; ARAN, f. 2, op. 1–50, d. 178, ll. 1–6a; f. 2, op. 1–50, d. 254, l. 43.

89. RGASPI, f. 17, op. 118, d. 970, ll. 221–222.

90. Ibid., ll. 221–222.

91. RGASPI, f. 17, op. 118, d. 989, ll. 116–141; d. 967, l. 188.

92. Leaders in the Central Committee routinely forwarded material concerning physiology to Zhdanov without comment. For instance, E. B. Babskii, a Ukrainian academician, sent Suslov a desperate note asking that the stenographic record of the session be edited to remove what he believed to be unjustified insults; Suslov wrote along the top of the telegram: "To Zhdanov. What's this about?" Likewise, when Orbeli or others approached anyone in the Party elite for support, their letters were immediately forwarded to Zhdanov. RGASPI, f. 17, op. 132, d. 347, ll. 95–101.

93. RGASPI, f. 17, op. 132, d. 348, ll. 259–261.

94. RGASPI, f. 17, op. 133, d. 180, l. 244.

95. RGASPI, f. 17, op. 119, d. 483, ll. 163–180.

96. RGASPI, f. 17, op. 133, d. 180, ll. 46–50.

97. RGASPI, f. 17, op. 133, d. 234, l. 63.

98. Ibid., ll. 39–60.

99. Ibid., ll. 60–63. The Armenian Party secretary proposed that Asratian be appointed director of the Institute of Physiology of Animals and Humans of the Armenian Academy of Sciences. Suslov suggested that Zhdanov talk to Asratian about the proposal, but Asratian refused to work in the Armenian academy. See ibid., l. 162.

100. Quoted in Sviderskii, *Orbeli*, 37.

101. Ibid., 121.

102. ARAN, f. 2, op. 1–51, d. 1851–1868; RGASPI, f. 17, op. 133, d. 180, ll. 6–28, 53–64.

103. RGASPI, f. 17, op. 133, d. 180, ll. 240–241.

104. Ibid., ll. 238–242.

105. Ibid., l. 29. Although the memo is not signed, the style and its location in the Agitprop papers suggest that Yuri Zhdanov was the author.

106. This meeting is covered in detail in Leibson, *Orbeli*, 117–149. See also RGASPI, f. 17, op. 133, d. 180, ll. 120–136.

107. Ibid.

108. Ibid.

109. Ibid.

110. Leibson reports that the cover letter was sent to Stalin through Aleksander Nikolaevich Nesmeianov, the president of the Academy of Sciences after Vavilov's death. Documents in the Party archive suggest, however, that Orbeli's cover letter was addressed to Aleksander Nikolaevich Poskrebyshev, Stalin's personal secretary. See Leibson, *Orbeli*, p. 156; RGASPI, f. 17, op. 133, d. 260, ll. 79–82.

111. RGASPI, f. 17, op. 133, d. 260, ll. 207–210.

112. Ibid., ll. 13–14.

113. Ibid., ll. 201–206. See also ll. 211–224 and Leibson, *Orbeli*, 157–177.

114. Iu. G. Murin, ed., *Iosif Stalin v ob"iatiiakh sem'i* (Moscow, 1993), 103. Svetlana Alliluyeva (translated by Priscilla Johnson McMillan), *Twenty Letters to a Friend* (New York, 1967), 211.

115. Leibson, *Orbeli*, 178–185; A. A. Zvorykin, ed., *Biograficheskii slovar' deiatelei estestoznaniia i tekhniki* (Moscow, 1959), 390.

CHAPTER 7
"EVERYONE IS WAITING"

1. ARAN, f. 1705, op. 1, d. 166, ll. 16–17.

2. This chapter examines the formulation of economic theory and its political and ideological implications, rather than economic history per se. The following authors provided helpful background: Richard B. Day, *The 'Crisis' and the 'Crash': Soviet Studies of the West* (London, 1981), and *Cold War Capitalism: The View from the West* (New York, 1995); Pekka Sutela, *Economic Thought and Economic Reform in the Soviet Union* (Cambridge, 1991), and *Socialism, Planning and Optimality: A Study in Soviet Economic Thought* (Helsinki, 1984); Werner Hahn, *Postwar Soviet Politics: The Fall of Zhdanov and the Defeat of Moderation, 1946–1953* (Ithaca, 1982); and Mark Harrison, *Soviet Planning in Peace and War, 1938–1945* (Cambridge, 1985).

3. For an important exception, see L. A. Openkin, "I. V. Stalin: poslednii prognoz budushchego," *Voprosy istorii KPSS* 1991 (7): 113–128.

4. *History of the Communist Party of the Soviet Union (Bolsheviks): Short Course* (New York, 1939). See also David Brandenberger, *National Bolshevism: Stalinist Mass Culture and the Formation of Russian National Identity, 1931–1956* (Cambridge, 2002).

5. RGASPI, f. 558, op. 11, d. 716, l. 22.

6. According to Stalin's secretary's notebook, the meeting took place at Stalin's Kremlin office from 1:20 p.m. to 2:50 p.m. on January 29, 1941. The four other economists were B. L. Markus, G. P. Kosiachenko, I. A. Trakhtenberg, and A. I. Pashkov. The secretary's list (published in *Istoricheskii arkhiv* 1996 [2]: 39) corresponds to the minutes found in ARAN, f. 1705, op. 1, d. 166, ll. 14–26.

7. ARAN, f. 1705, op. 1, d. 166, ll. 17–22.

8. Ibid., ll. 22–24.

9. Ibid., ll. 16–17.

10. Ibid., l. 21.

11. Ibid., ll. 14–26.

12. ARAN, f. 1723, op. 1, d. 32, ll. 2–3.

13. Harrision, *Soviet Planning*, 224–230; Raya Dunayevskaya, "A New Revision of Marxian Economic," *American Economic Review* 1944 (3): 536–537. (Pages 501–530 of the journal contain translations of the editorial from *Under the Banner of Marxism*.) See "Nekotorye voprosy prepodavaniia politicheskoi ekonomii," *Pod znamenem marksizma* 1943 (7–8): 56–78.

14. RGASPI, f. 17, op. 125, d. 307, ll. 78–89.

15. Ibid., ll. 155–161.

16. RGASPI, f. 17, op. 125, d. 380, ll. 1–4; RGASPI, f. 17, op. 125, d. 381.

17. RGASPI, f. 558, op. 11, d. 1226, l. 18.

18. RGASPI, f. 17, op. 121, d. 643, ll. 26–28.

19. ARAN, f. 9396, op. 1, d. 123, l. 71; RGASPI, f. 17, op. 118, d. 295, ll. 26–41.

20. Day, *Cold War Capitalism*, 45–48; RGASPI, f. 17, op. 125, d. 551, ll. 90–97.

21. Yoram Gorlizki and Oles Khlevniuk, *Cold Peace: Stalin and the Ruling Circle, 1945–1953* (Oxford, 2004), 80–89; Ethan Pollock, "The Politics of Knowledge: Party Ideology and Soviet Science, 1945–1953," Ph.D. dissertation, University of California, Berkeley, 2000, 406–409.

22. ARAN, f. 615, op. 1, d. 5, ll. 19–26.

23. RGASPI, f. 17, op. 132, d. 158, l. 75; RGASPI, f. 17, op. 118, d. 477, ll. 76–77. The fact that the "other nationalities" were not identified suggests that the Central Committee was interested specifically in Jews.

24. RGASPI, f. 17, op. 133, d. 10, l. 8.

25. RGASPI, f. 17, op. 132, d. 343, ll. 60–62.

26. RGASPI, f. 588, op. 11, d. 1226, l. 29.

27. Again the participants identified in the minutes correspond to the list of visitors to Stalin's office. See *Istoricheskii arkhiv* 1997 (1): 12. Pavel Iudin was a Soviet philosopher who began an ascent to the top of the party apparatus after Aleksandrov's dismissal from Agitprop and Zhdanov's death. He was appointed as a candidate member of the Politburo in 1952.

28. In 1949 *Voprosy ekonomiki* published articles on the economies of capitalist countries imbued with ideological spirit, if not economic substance: "The Currency Crisis and Inflation in the Capitalist World," "Currency Crisis in England," "The Expansion of the U.S. Oil Monopoly in South America," "Capitalist Finance: A Weapon of Imperialist Aggression," "Economic Contradictions in the Battle between American and English Imperialism," "The Imperialist Struggle for Antarctica," "Bourgeois Cosmopolitanism in the Service of American Imperialists," etc.

29. RGASPI, f. 17, op. 133, d. 41, ll. 5–6.

30. Ibid., ll. 1–4. The other members of the authors' collective were V. P. D"iachenko, deputy director of the Institute of Economics and assistant editor of *Questions of Economics*, who was an expert on finance and credit in the USSR; A. M. Pashkov, assistant editor of *Questions of Economics*; L. M. Gatovskii, specialist on the Soviet economy from the Higher Party School and the Department of Socialist Political Economy of the Institute of Economics; G. M. Kozlov, a teacher of political economy at the Frunze Military Academy; M. D. Laptev, a member of Lysenko's Lenin All-Union Agricultural Academy whose *Pravda* article in 1948 harshly criticized Varga; A. A. Pal'tsev, chair of the Department of Political Economy of the Moscow Institute of Soviet Cooperative Trade; A. P. Liapin, who had written an article in *Questions of Economics* in 1949 on Stalin and the road to communism in the USSR; P. K. Vasutin; and N. F. Makarova.

31. RGASPI, f. 17, op. 118, d. 747, ll. 91–93, 132. The decrees were written on February 23 and officially approved at a meeting of the Secretariat on March 4. Iudin's leave of absence, which lasted two months longer than the others, needed approval from the Poliburo.

32. RGASPI, f. 17, op. 132, d. 343, ll. 1–16.

33. *Istoricheskii arkhiv* 1997 (1): 12.

34. RGASPI, f. 17, op. 133, d. 41, ll. 18–19.

35. Ibid., ll. 19–20, 24–25.

36. Ibid., ll. 20–25.

37. Ibid., ll. 24.

38. Ibid., ll. 15–16.

39. Ibid., ll. 8–17.

40. Ibid., ll. 17.

41. Ibid., ll. 38–69, 142–143.

42. RGASPI, f. 17, op. 133, d. 35, ll. 1–19.

43. RGASPI, f. 17, op. 41, l. 150.

44. RGASPI, f. 17, op. 119, d. 206, ll. 51, 173–176; RGASPI, f. 17, op. 133, d. 41, ll. 87–88, 142–143.

45. ARAN, f. 1767, op. 1, d. 145, l. 25.

46. Ibid.

47. RGASPI, f. 17, op. 133, d. 42, ll. 178–179; RGASPI, f. 17, op. 3, d. 1091, ll. 33–34.

48. RGASPI, f. 17, op. 133, d. 42, ll. 193–204.

49. Unfortunately, I have been unable to locate any evidence indicating why the meeting was delayed from September to October and then from October to November.

50. Until the opening of the Soviet archives, the content of the political economy discussion was completely unknown to historians. Despite the discussion's significance for Stalin's last theoretical work and its usefulness as a lens for the study of Marxist-Leninist theory in the last years of the Stalin regime, the stenographic records of the meetings have barely come to the attention of historians. As in the previous chapters, my analysis offers a synopsis of the content of the discussion as culled from stenographic reports. In this chapter, the process is complicated by two factors. First, the meeting lasted more than twice as long as any of the previous postwar discussions. Second, unlike the philosophy, biology, linguistics, and physiology meetings, no published record of the political economy meeting exists. Thus all the information on the content of the political economy meeting was obtained from over four thousand pages of archival documents.

51. RGASPI, f. 17, op. 133, d. 85, l. 161.

52. RGASPI, f. 17, op. 133, d. 45, l. 33.

53. RGASPI, f. 17, op. 133, d. 56, l. 163.

54. RGASPI, f. 17, op. 133, d. 70, l. 102.

55. RGASPI, f. 17, op. 133, d. 56, l. 90.

56. RGASPI, f. 17, op. 133, d. 62, l. 7.

57. Ibid., l. 138.

58. RGASPI, f. 17, op. 133, d. 70, ll. 1–13.

59. RGASPI, f. 17, op. 133, d. 59, ll. 68–100.

60. RGASPI, f. 17, op. 133, d. 48, l. 145.

61. RGASPI, f. 17, op. 133, d. 56, ll. 41–88.

62. RGASPI, f. 17, op. 133, d. 45, ll. 189–226.

63. RGASPI, f. 17, op. 133, d. 79, ll. 40–41.

64. RGASPI, f. 17, op. 133, d. 45, l. 118; RGASPI, f. 17, op. 133, d. 56, l. 150; RGASPI, f. 17, op. 133, d. 76, l. 74; RGASPI, f. 17, op. 133, d. 79, ll. 147–171; RGASPI, f. 17, op. 133, d. 82, l. 36; RGASPI, f. 17, op. 133, d. 88, ll. 2–33.

65. RGASPI, f. 17, op. 133, d. 56, ll. 63–64.

66. RGASPI, f. 17, op. 133, d. 45, ll. 21, 24; RGASPI, f. 17, op. 133, d. 56, ll. 190–203.

67. RGASPI, f. 17, op. 133, d. 56, l. 84; RGASPI, f. 17, op. 133, d. 59, ll. 2–3, 71; RGASPI, f. 17, op. 133, d. 62, ll. 7, 160; RGASPI, f. 17, op. 133, d. 66, l. 43; RGASPI, f. 17, op. 133, d. 70, ll. 103–106, 163; RGASPI, f. 17, op. 133, d. 73, ll. 164, 190; RGASPI, f. 17, op. 133, d. 82, l. 170.

68. Not every issue concerning modern capitalism was controversial. When P. M. Pal'tsev, from the Moscow Cooperative Institute, insisted that the textbook address the category of national income and modern capitalism, the authors readily capitulated, admitting that they had erred in not including the subject originally. Likewise, there was no argument that the textbook should provide a critique of the nationalization of industry that was taking place in bourgeois countries such as England. See RGASPI, f. 17, op. 133, d. 48, ll. 2–4; RGASPI, f. 17, op. 133, d. 59, l. 138; and RGASPI, f. 17, op. 133, d. 100, l. 74.

69. RGASPI, f. 17, op. 133, d. 45, l. 89.

70. RGASPI, f. 17, op. 133, d. 48, ll. 188–212.

71. RGASPI, f. 17, op. 133, d. 59, l. 101; RGASPI, f. 17, op. 133, d. 62, l. 100; RGASPI, f. 17, op. 133, d. 79, l. 121–146.

72. RGASPI, f. 17, op. 133, d. 82, l. 9, 118; RGASPI, f. 17, op. 133, d. 94, l. 67. Varga's notes in preparation for the meeting are available in ARAN, f. 1513, op. 1, d. 61.

73. RGASPI, f. 17, op. 133, d.140 ll. 42–97; RGASPI, f. 17, op. 133, d. 143, ll. 1–103.

74. RGASPI, f. 17, op. 133, d. 45, ll. 82–84; RGASPI, f. 17, op. 133, d. 48, ll. 69, 114.

75. RGASPI, f. 17, op. 133, d. 59, l. 3; RGASPI, f. 17, op. 133, d. 62, ll. 165, 197; RGASPI, f. 17, op. 133, d. 76, l. 164; RGASPI, f. 17, op. 133, d. 79, ll. 32–33.

76. RGASPI, f. 17, op. 133, d. 85, l. 19.

77. RGASPI, f. 17, op. 133, d. 59, ll. 31–67.

78. RGASPI, f. 17, op. 133, d. 76, l. 91; RGASPI, f. 17, op. 133, d. 56, l. 162.

79. RGASPI, f. 17, op. 133, d. 48, ll. 45–54, 114; RGASPI, f. 17, op. 133, d. 56, ll. 90–120; RGASPI, f. 17, op. 133, d. 59, l. 126; RGASPI, f. 17, op. 133, d. 66, l. 81; RGASPI, f. 17, op. 133, d. 73, l. 29; RGASPI, f. 17, op. 133, d. 79, l. 226; RGASPI, f. 17, op. 133, d. 85, l. 180.

80. RGASPI, f. 17, op. 133, d. 45, ll. 162–164; RGASPI, f. 17, op. 133, d. 94, l. 106.

81. RGASPI, f. 17, op. 133, d. 59, l. 128; RGASPI, f. 17, op. 133, d. 66, l. 20; RGASPI, f. 17, op. 133, d. 82, l. 147; RGASPI, f. 17, op. 133, d. 106, l. 43.

82. RGASPI, f. 17, op. 133, d. 82, l. 24; RGASPI, f. 17, op. 133, d. 88, l. 2; RGASPI, f. 17, op. 133, d. 97, ll. 23, 80.

83. Openkin, "I. V. Stalin," 116.

84. RGASPI, f. 17, op. 133, d. 100, ll. 46–79.

85. Ibid., ll. 79–94.

86. RGASPI, f. 17, op. 133, d. 103, ll. 1–50.

87. RGASPI, f. 17, op. 133, d. 106, ll. 42–86.

88. Ibid.

89. RGASPI, f. 558, op. 11, d. 1248, ll. 1–3.

90. RGASPI, f. 17, op. 133, d. 136; RGASPI, f. 17, op. 133, d. 138; RGASPI, f. 17, op. 133, d. 140; RGASPI, f. 17, op. 133, d. 143; RGASPI, f. 17, op. 133, d. 148; RGASPI, f. 17, op. 133, d. 151; RGASPI, f. 17, op. 133, d. 155; RGASPI, f. 17, op. 133, d. 157; RGASPI, f. 17, op. 133, d. 159.

91. RGASPI, f. 17, op. 133, d. 42, ll. 217–227; RGASPI, f. 17, op. 133, d. 41, ll. 161–163; ARAN, f. 1723, op. 1, d. 32, l. 5.

92. RGASPI, f. 17, op. 133, d. 42, ll. 217–227; RGASPI, f. 17, op. 133, d. 164, and RGASPI, f. 17, op. 133, d. 165.

93. RGASPI, f. 83, op. 1, d. 8, ll. 19–77.

94. RGASPI, f. 17, op. 133, d. 216, ll. 1–56; RGASPI, f. 17, op. 132, 503, ll. 211–245; RGASPI, f. 17, op. 133, d. 163, ll. 33–39.

95. RGASPI, f. 588, op. 11, d. 1242.

96. RGASPI, f. 588, op. 11, d. 1248, ll. 34–42.

97. J. Stalin, *Economic Problems of Socialism in the USSR* (Moscow, 1952), 5–13. Quotations are taken from the English translations provided by the Foreign Language Publishing House in Moscow. For the original, see I. Stalin, *Ekonomicheskie problemy sotsializma v SSSR* (Moscow, 1952).

98. Stalin, *Economic Problems*, 42–46.

99. Ibid., 14.

100. Ibid., 13–29.

101. Ibid., 29–34.

102. Ibid., 34–37.

103. Ibid., 37–41.

104. Ibid., 50–53.

105. ARAN, f. 1705, op. 1, d. 166, ll. 54–56.

106. RGASPI, f. 17, op. 133, d. 216, ll. 148–149.

107. RGASPI, f. 17, op. 133, d. 215, l. 1. The economists were Gatovskii, Kuz'minov, Laptev, Leont'ev, Ostrovitianov, Pereslegin, Pashkov, Shepilov, Iudin, Atlas, Arakelian, Bolgov, Vasil'eva, Guskaov, Kozlov, Lubimov, and Rubinshtein.

108. RGASPI, f. 17, op. 133, d. 215, l. 2.

109. Ibid., ll. 3–4.

110. Ibid., ll. 4–5.

111. Ibid., l. 6.

112. Ibid., ll. 9–10.

113. Ibid., ll. 9–13.

114. RGASPI, f. 17, op. 3, d. 1092, l. 96.

115. Stalin's responses were sent to select economists upon completion, confirming the dates provided in the published versions. See ARAN, f. 1767, op. 1, d. 146, ll. 129, 153.

116. Stalin, *Economic Problems*, 54. Notkin's original letter is not included; therefore its content must be extrapolated from Stalin's response.

117. Stalin, *Economic Problems*, 55–56.

118. Ibid., 62–63.

119. RGASPI, f. 17, op. 133, d. 217, ll. 32–101.

120. RGASPI, f. 17, op. 133, d. 94, ll. 2–40; RGASPI, f. 17, op. 133, d. 153, ll. 6–8; RGASPI, f. 17, op. 133, d. 159, ll. 23–27.

121. RGASPI, f. 17, op. 133, d. 216, ll. 10–18, 56.

122. RGASPI, f. 17, op. 133, d. 302, l. 62.

123. RGASPI, f. 17, op. 133, d. 217, ll. 27–31, 133–138, 146–165.

124. RGASPI, f. 17, op. 133, d. 217, ll. 151–171.

125. RGASPI, f. 83, op. 1, d. 7, ll. 78–79.

126. Iaroshenko's story corresponds to the available archival records. See Iaroshenko, "Kak ia stal 'poslednei zhertvoi'," *Pravda*, September 29, 1989.

127. Stalin, *Economic Problems*, 96–104.

128. Ibid., 94.

129. *Pravda*, October 14, 1952 quoted in Openkin, "I. V. Stalin," 113–114.

130. RGASPI, f. 17, op. 133, d. 282, ll. 11–15.

131. ARAN, f. 1705, op. 1, d. 175, ll. 1–82; RGASPI, f. 17, op. 133, d. 302, ll. 10, 72–73.

132. RGASPI, f. 83, op. 1, d. 8, ll. 119–120. V. Ostrovitianov et. al., *Politicheskaia ekonomiia: uchebnik* (Moscow, 1954).

133. Ostrovitianov et al., *Politicheskaia ekonomiia*, 4.

CHAPTER 8
CONCLUSION

1. Alexander Vucinich, *Empire of Knowledge* (Berkeley, 1984), 245–246. The original speech, along with Stalin's corrections of preliminary drafts, is located in RGASPI, f. 592, op.1, d. 6, ll. 67–68.

2. RGASPI, f. 17, op. 133, d. 4, ll. 60–68.

3. Ibid., ll. 69–70.

4. For a different opinion, see Nikolai Krementsov, *Stalinist Science* (Princeton, 1997), 159.

5. The significance of ideology in the Cold War became all the more clear in the 1990s: the conflict ended when elite members of the Communist Party lost their faith in the system, not when the country's economy faltered—which happened in the 1970s—or when its militarily was thoroughly defeated—which, of course, never happened. For an elaboration of the ideological dimensions to the end of the Cold War, see Stephen Kotkin, *Armageddon Averted: The Soviet Collapse, 1970–2000* (Oxford, 2001).

6. See Dmitrii Volkogonov, *Stalin, politicheskii portret, kniga 2* (Moscow, 1996), 555–558.

7. I. N. Zhukov, "Bor'ba za vlast' v rukovodsve SSSR v 1945–1952 godakh," *Voprosy istorii* 1995 (1): 23–29. Yoram Gorlizki and Oleg Khlevniuk argue that the whole point of the 1951 political economy discussion was to offer Stalin a pretext to draft a report to the upcoming Party congress. See their *Cold Peace: Stalin and the Ruling Cirle, 1945–1953* (Oxford, 2004), 145–146.

8. Robert Tucker, *Stalin as Revolutionary: A Study in History and Personality, 1879–1929* (New York, 1973), 319.

9. We might think of "coryphaeus of science" as the intellectual counterpart to Stalin's cultural, social, and political identities as outlined in Alfred J. Rieber, "Stalin, Man of the Borderlands," *American Historical Review* 106 (2001): 1651–1691.

10. This point was first noted in David Joravsky, *The Lysenko Affair* (Cambridge, 1970), 153–157, and in Alexander Vucinich, *Empire of Knowledge: The Academy of Sciences of the USSR (1917–1970)* (Berkeley, 1984), 252–254, and has since been supported by archival evidence in the Party archives. For details, see RGASPI, f. 17, op. 119, d. 817, d. 851, d. 1036, d. 1037, and d. 1038.

11. "Nel'zia peredelyvat' zakony prirody," *Izvestiia TsK KPSS* 1991 (2): 104–110.

12. For excellent examples of how Khrushchev-era policies envisioned a rational science free from economic and ideological constraints, see Paul R. Josephson, *New*

Atlantis Revisited: Akademgorodok, the Siberian City of Science (Princeton, 1997), and Slava Gerovitch, *From Newspeak to Cyberspeak: A History of Soviet Cybernetics* (Cambridge, 2002.)

13. Quoted in Stephen Fortescue, *The Communist Party and Soviet Science* (Baltimore, 1986), 19.

14. Matthew Evangelista, *Unarmed Forces: The Transnational Movement to End the Cold War* (Ithaca, 1999), and David Holloway, *Stalin and the Bomb* (New Haven, 1994).

15. Quoted in Robert D. English, *Russia and the Idea of the West: Gorbachev, Intellectuals and the End of the Cold War* (New York, 2000), 76.

16. For a detailed and fascinating account of this process in cybernetics, see "Cybernetics in Rebellion," in Gerovitch, *From Newspeak to Cyberspeak.*

17. Andrei Sakharov, *Mir, progress, prava cheloveka: stat'i i vystupleniia* (Leningrad: 1990), 54.

18. Andrei Sakharov (translated by the *New York Times*), *Progress, Peaceful Coexistence, and Intellectual Freedom* (New York, 1968), 1. Not all dissidents attacked the system for not living up to its scientific claims. Alexander Solzhenitsyn, for instance, saw the problem in almost opposite terms. The problem was too much rationality, or as he put it: "The endless, infinite progress dinned into our heads by the dreamers of the Enlightenment cannot be accomplished." Alexander Solzhenitsyn, *Letter to the Soviet Leaders* (New York, 1974), 21.

19. Quoted in English, *Russia and the Idea of the West*, 200.

20. On the degree to which Gorbachev's reforms laid bare the decaying ideology, as it had been formulated since Stalin's articles on linguistics, see Alexei Yurchak, "Soviet Hegemony of Form: Everything Was Forever, Until It Was No More," *Comparative Studies in Society and History* 45 (2003): 480–510.

21. John Cornwell, *Hitler's Scientists: Science, War and the Devil's Pact* (New York, 2003), 21–37.

22. Robert N. Proctor, *Racial Hygiene: Medicine under the Nazis* (Cambridge, 1988), 50, 295.

23. Cathryn Carson, "New Models for Science in Politics: Heisenberg in West Germany," *Historical Studies in the Physical and Biological Sciences* 1999 (1): 115–172.

24. Quoted in James H. Williams, "Fang Lizhi's Big Bang: A Physicist and the State in China," *Historical Studies in the Physical and Biological Sciences* 1999 (1): 56–57.

25. H. Lyman Miller, *Science and Dissent in Post-Mao China: The Politics of Knowledge* (Seattle, 1996).

26. David A. Hollinger, "Free Enterprise and Free Inquiry: The Emergence of Laissez-Faire Communitarianism in the Ideology of Science in the United States," *New Literary History* 21 (1990): 897–917.

27. *Scientific Integrity in Policymaking: An Investigation into the Bush Administration's Misuse of Science*, a report by the Union of Concerned Scientists (February 2004).

28. Loren R. Graham, *What Have We Learned about Science and Technology from the Russian Experience?* (Stanford, 1998), 52–73.

29. See Douglas R. Weiner, *A Little Corner of Freedom: Russian Nature Protection from Stalin to Gorbachev* (Berkeley, 1999), and Holloway, *Stalin and the Bomb.*

BIOGRAPHICAL NOTES

Aleksandrov, Georgii Fedorovich (1908–1961). Philosopher and Party figure; joined the Communist Party in 1928. Graduated from the Institute of History and Philosophy in 1932, received his doctorate in philosophy in 1939, and became a member of the Soviet Academy of Sciences in 1946. He was deputy head and then head of the Agitation and Propaganda Section (1939–1947) while also directing the Central Committee's Higher Party School (1939–1946). After the Central Committee–sponsored discussion of his book *History of Western European Philosophy* (for which he had won a Stalin Prize in 1946), he was removed from his position in Agitprop and appointed director of the Institute of Philosophy (1947–1954).

Beria, Lavrenty Pavlovich (1899–1953). Party and government figure; joined the Communist Party in 1917. After serving as the first secretary of the Georgian Communist Party, he became deputy head and then the head of Stalin's secret police (People's Commissariat of Internal Affairs, or NKVD) (1938–1945), deputy head of the Council of Ministers (1945–1953), member of the Politburo and presidium (1945–1953), and chair of the Special Committee on the atomic bomb (1945–1953).

Bykov, Konstantin Mikhailovich (1886–1959). Physiologist. Graduated from Kazan University in 1912. From 1921 to 1950 he worked at the Institute of Experimental Medicine—the first eleven years under Ivan Pavlov. He became a member of the Academy of Medical Sciences in 1944 and a member of the Soviet Academy of Sciences in 1946. From 1948 to 1950 he directed the academy's Institute of Physiology of the Central Nervous System. In the aftermath of the Pavlov Session, he became the director of the academy's Institute of Physiology.

Chikobova, Arnol'd Stepanovich (1898–1985). Linguist. Graduated from the University of Tbilisi in 1922 and then taught there (1926–1985) and headed

its Department of Caucasian Languages (1933–1960). At the Institute of Linguistics of the Georgian Academy of Sciences he chaired the Department of
Iberian-Caucasian Languages (1936–1985) and directed the institute (1950–
1952). In 1941 he became a member of the Georgian Academy of Sciences.

Kaftanov, Sergei Vasil'evich (1905–1978). Chemist and government figure;
joined the Communist Party in 1926. Graduated from the Mendeleev Chemi-
cal-Technical Institute in Moscow in 1931 and began working in the Central
Committee apparatus in 1937 as the head of the All-Union Committee for
Higher Education. He was a representative of the State Defense Committee
for science (1941–1945) and minister of higher education (1946–1951).

Kapitsa, Peter Leonidovich (1894–1984). Physicist. Graduated from the Petrograd Polytechnical University in 1918. Worked in the Cavendish Laboratory in Cambridge, England, from 1921 until 1934, when he was forced to
stay in the USSR. In Moscow, he set up and directed the Institute for Physical
Problems (1935–1946, 1955–1984). He was awarded two Stalin Prizes (1941
and 1943), named Hero of Socialist Labor (1945), and received the Nobel
Prize in Physics (1978).

Kedrov, Bonifatii Mikhailovich (1903–1983). Philosopher; joined the Communist Party in 1918. A graduate of the Chemistry Department of Moscow
State University (1930), he was a leading philosopher of science at the Institute of Philosophy (1945–1949), the chief editor of the journal *Questions of
Philosophy* (1947–1949), and on the editorial board of the *Great Soviet Encyclopedia* (1949–1952).

Kurchatov, Igor Vasil'evich (1903–1960). Physicist; joined the Communist
Party in 1948. After graduating from the Crimean University (1923), he went
to work at the Leningrad Physical-Technical Institute under the direction
of Abram Ioffe. In 1943 he became a member of the Academy of Sciences
and the scientific director of the Soviet atomic project, a position he held
until 1960. He was named a Hero of Socialist Labor three times (1949, 1951,
and 1954).

Leont'ev, Lev Abramovich (1901–1974). Economist; joined the Communist
Party in 1919. Graduated from the Institute of Red Professors (1925) and
worked as the head of the authors' collective responsible for writing the Central Committee–commissioned textbook on political economy (1937–1954).

Lysenko, Trofim Denisovich (1898–1976). Agronomist. After graduating
from the Kiev Agricultural Institute in 1925, he worked on plant breeding in
Azerbaijan and Ukraine. He became a member of the Ukrainian Academy of
Sciences in 1934, a member of the Lenin All-Union Agricultural Academy
in 1935, and a member of the Soviet Academy of Sciences in 1939. He directed the Agricultural Academy (1938–1956, 1961–1962) and the Genetics

Department of the Timiriazev Agricultural Academy (1948–1965). He was awarded the Stalin Prize three times (1941, 1943, and 1949), Hero of Socialist Labor once (1945), and served as the deputy head of the USSR Supreme Soviet (1938–1956).

Maksimov, Aleksander Aleksandrovich (1891–?). Philosopher; joined the Communist Party in 1918. After graduating from the Physics and Mathematics Department of the University of Kazan in 1916, he turned his attention to the philosophy of science. Beginning in 1929 he taught at the Institute of Red Professors and then in the Philosophy Department at Moscow State University. He was a researcher at the Institute of Philosophy of the Academy of Sciences and a member of the editorial board of *Questions of Philosophy*.

Malenkov, Georgii Maksimilianovich (1902–1988). Party and government figure; joined the Communist Party in 1920. After graduating from the Bauman Higher Technical College in Moscow in 1925, he joined the Central Committee apparatus and then the Moscow Party Committee. He served as head of the section of higher Party organs (1934–1939), secretary of the Central Committee (1939–1946, 1948–1953), head of the Cadres Administration (1939–1946), a member of the State Defense Committee (1941–1945), deputy head of the Council of People's Commissars (1944–1946), and deputy head of the Council of Ministers (1946–1953). He was named a Hero of Socialist Labor in 1943.

Marr, Nikolia Iakovlevich (1865–1934). Linguist. Graduated from the Department of Oriental Languages of the University of St. Petersburg in 1890; became dean of the faculty of Oriental languages there in 1911 and a member of the Russian Academy of Sciences in 1912. As one of the few members of the Imperial Academy of Sciences to endorse the Revolution, he won administrative support for his linguistics theories. He was director of the Institute for Japhetic Languages of the Soviet Academy of Sciences (1921–1934) and vice president of the Academy of Sciences (1930–1934).

Meshchaninov, Ivan Ivanovich (1883–1957). Linguist and archaeologist. Graduated from the University of St. Petersburg in 1907 and from the Institute of Archaeology in 1910. A student of Nikolai Marr's, he directed the Institute of Anthropology, Archaeology, and Ethnography (1933–1937) and the Institute of Language and Thought (1933–1950), both of the Academy of Sciences. He served as the academic secretary of the academy's Division for Literature and Linguistics (1934–1950) and won two Stalin Prizes (1943, 1946).

Michurin, Ivan Valdimirovich (1855–1935). Horticulturalist. Without any formal education, he relied on crude hybridization in an unsuccessful attempt to develop new plant varieties. In the 1920s he won support from the Soviet government, which made him a popular hero as a practical breeder who eschewed the pretensions of science. He became an honorary member of the Academy of Sciences in 1935.

Mitin, Mark Borisovich (1901–1987). Philosopher; joined the Communist Party in 1919. Graduate of the Institute of Red Professors (1929), member of the Central Committee of the Communist Party (1939–1944, 1950–1956), director of the Institute of Marxism-Leninism (1939–1944), chief editor of *Under the Banner of Marxism* (1939–1944), and then an editor of the *Literary Gazette* in the late 1940s and early 1950s.

Orbeli, Leon Abgarovich (1882–1958). Physiologist. After graduating from the Military Medical Academy in 1904, he worked from 1907 to 1921 under Ivan Pavlov in the Physiology Department of the Institute of Experimental Medicine. In addition to teaching and other posts, he was the director of the Military Medical Academy (1943–1950), director of the Physiology Institute of the Academy of Sciences (1936–1950), director of the Institute of Physiology and Pathology of the Higher Nervous System (1936–1950), academic secretary of the Biology Division of the Academy of Sciences (1939–1948), and vice president of the Academy of Sciences (1942–1946). He became a member of the Academy of Sciences (1935), the Armenian Academy of Sciences (1943), and the Academy of Medical Sciences (1944), received a Stalin Prize (1941), and was named Hero of Socialist Labor (1945).

Ostrovitianov, Konstantin Vasil'evich (1892–1969). Economist and Party figure; joined the Communist Party in 1914. Graduated from the Moscow Commercial Institute in 1917 and served as secretary of the Zamoskvorech'e Military Revolutionary Committee during the October Revolution. Director of the Institute of Economics of the Soviet Academy of Sciences (1947–1953), chief editor of *Questions of Economics* (1948–1954), acting academic secretary of the Division for Economics, Philosophy, and Law of the Academy of Sciences (1949–1953), and named a candidate member of the Central Committee of the Communist Party (1952). He was named a corresponding member of the Academy of Sciences in 1939 and a full member in 1953. Along with Lev Leont'ev, he was a main editor of the Central Committee–sponsored textbook on political economy.

Pavlov, Ivan Petrovich (1849–1936). Physiologist. Graduated from St. Petersburg University (1875) and the Medical Surgical Academy (1879) and became a member of the Russian Academy of Sciences in 1907. He was director of the Physiology Department of the Medical Surgical Academy (1895–1924), the Institute of Experimental Medicine (1891–1936), and the Physiology Institute of the Soviet Academy of Sciences (1925–1936). He won a Nobel Prize in 1904.

Shepilov, Dmitrii Trofimovich (1905–1995). Economist and Party figure; joined the Communist Party in 1926. Graduated from the Social Sciences College of Moscow State University in 1926 and the Agrarian Institute of Red Professors in 1933. Served as deputy head and then head of the Central Committee's section on agricultural sciences (1935–1941), as an editor of *Pravda* (1946–1947), as deputy head and then head of Agitprop (1947–1949), and as a Central Committee inspector (1950–1953).

Suslov, Mikhael Andreevich (1902–1982). Party figure; joined the Communist Party in 1921. Member of the Central Committee (1941–1982), member of the Orgburo (1946–1952), a secretary of the Central Committee (1947–1982), and chief editor of *Pravda* (1949–1951).

Vavilov, Sergei Ivanovich (1891–1951). Physicist. Graduated from the Physical-Mathematical College of the Moscow State University in 1914, where he taught from 1918 to 1932. He was the scientific director of the State Optical Institute (1932–1945), a full member of the Academy of Sciences (1932), and president of the Academy of Sciences (1945–1951). He received the Stain Prize twice (1946 and posthumously in 1951).

Voznesenskii, Nikolai Alekseevich (1903–1950). Party and government figure, economist; joined the Communist Party in 1919. Graduated from the Sverdlov Communist University in 1924 and the Economic Institute of Red Professors in 1931. He received a doctorate in 1935 and became a member of the Academy of Sciences in 1943. He served as deputy head and then head of the State Planning Commission (1937–1949), as a member of the State Defense Committee (1942–1945), and as deputy and first deputy chair of the Council of People's Commissars and then Council of Ministers (1939–1949). Candidate and then full member of the Politburo (1941–1949). Arrested in October 1949 and shot in 1950.

Zhdanov, Andrei Aleksandrovich (1896–1948). Party and government figure; joined the Communist Party in 1915. In 1934, after serving as a regional Party leader in Tver and Nizhnii Novgorod, he became secretary of the Leningrad Party Committee, where he served until 1944, secretary of the Central Committee, and candidate and then full member of the Politubro (1935 to 1948).

Zhdanov, Yuri Andreievich (b. 1919). Party figure, son of Andrei Zhdanov; joined the Communist Party in 1944. After graduating from Moscow State University with a degree in chemistry in 1941 and then studying the philosophy of science, he served as the head of the Central Committee's Science Section from 1947 to 1954. In 1949 he married Stalin's daughter, Svetlana.

ACKNOWLEDGMENTS

After I received my bachelor's degree from Tufts University in 1991 my adviser Martin Sherwin arranged for me to go to Moscow to live and work. I had taken Russian language courses in college and had been to the Soviet Union the year before for a semester of study, but I was not at all certain what I was getting myself into. As the coup against Gorbachev failed, the Soviet Union collapsed, and political, economic, and cultural revolutions went on around me, I got hooked on trying to figure out Russia. It has been a winding path from that moment to this book, and it brings me great pleasure to be able to thank those who have aided me along the way. In the early 1990s Natalia Tarasova provided me with a safe haven at the Mendeleev University and introduced me to many members of the Russian scientific establishment. Sergei Cherepennikov, Sasha Radostev, Marc Kasher, and others helped make Moscow my home for almost two years. But it was Marty who guided me back to history, first by exposing me to the exciting work being done at the time on the history of the Soviet atomic bomb project and then by strongly encouraging me to go to the University of California for graduate school.

At first blush, the Bay Area, with its sun, vistas, and cafés, hardly seemed to me like a place conducive to the study of the USSR. Then I got to know Reggie Zelnik and Yuri Slezkine. Russians sometimes talk about the value of *obshchenie* (roughly, interpersonal contact) as a key element of a meaningful life. In formal and informal settings Reggie and Yuri made sure that this feeling was alive and well in Berkeley. Without them as mentors I never would have remained in academia or seen this book, or the dissertation it began as, through to completion. Yuri's healthy appetite for adventure lured me in even more than his remarkable ideas and intellect. Reggie always offered sage advice about how to approach both history and life. It meant the world to know that behind his words stood his shining example. David Holloway of Stanford University generously agreed to be on my dissertation committee and graciously shared his expertise on postwar Soviet science. Victoria Bonnell provided kind support and critical comments when I needed them most. Alexander Vucinich consistently took time away from his own work to talk with me about Soviet science and never made me feel small when I discovered things that he knew all along. Cathryn Carson and David Hollinger helped with questions about the history of science, and Gregory Grossman guided me through some of the finer points of Soviet economic theory. Fellow graduate students and friends in Berkeley in history and in other fields, including Chad Arnold, Dan Covitz, Paul Davies, Durba Ghosh, Mark Glickman, Tracy Gordon, Susanne Kauer, Marcy Norton, Christian Redfearn, Paul Sabin, Doug Shoemaker, and Helen Sillett, provided unparalleled camaraderie and intellectual stimulation.

This book is based on archival research in Russia between 1997 and 2002. At an early stage, Alexei Kojevnikov encouraged me to take on this project and helped steer me in the right direction in the archives. Generous grants from International Research and Exchange Board (IREX) and Fulbright-Hays provided the funding for most of the research. The Social Science Research Council (SSRC) awarded me a grant to write the dissertation. I am deeply indebted to the many archivists at RGASPI, ARAN, GARF, and TsAODM and to the librarians at UC Berkeley and the State Historical Library in Moscow who helped me locate relevant material. Vladimir Esakov and Elena Levina welcomed me in their home and dacha and answered questions about Stalin's science policy and the Soviet bureaucracy. Valdimir Vizgin and others at the Academy Sciences Institute for the History of Science and Technology provided me with an institutional base of support and hours of advice and guidance. D'Ann Penner encouraged me to travel to Rostov to interview Yuri Zhdanov and then took the time to show me the beauty of that part of Russia. More recently, Marina Dobronovskaya provided essential research assistance in locating relevant images for the book.

Postdoctoral fellowships at the George Washington University Center for History of Recent Science and the Harriman Institute at Columbia University provided me with the time and resources to conduct a follow-up research trip to Russia and to revise the manuscript. I would like to thank Horace Freeland Judson for teaching me many lessons on writing and all of my fellow postdocs for sharing their work and helping me with mine. I have been very fortunate that my nomadic existence has come to a halt at Syracuse University, where the Department of History has been warmly welcoming and where colleagues across the campus have been a pleasure to get to know. I am grateful to Samantha Herrick and Norman Kutcher for dedicating countless hours to discussing issues big and small related to this book and related to life.

Many people offered helpful comments and suggestions after reading individual chapters of this manuscript or the whole work in progress, including Andy Day, Paula Devos, Victoria Frede, Jennifer Gold, Gerald Greenberg,

Norman Kutcher, Fredrick Marquardt, Rebecca Balmas Neary, Marcy Norton, Paul Sabin, Olga Shevchenko, Katrina Shwartz, and Kiril Tomoff. Douglas Weiner and Loren Graham offered helpful suggestions for turning the dissertation into a book. Vladislav Zubok and an anonymous reader for Princeton University Press helped me tighten the argument and broaden its scope. Laura Engelstein, Louise McReynolds, and Daniel Orlovsky stepped in with sound and heartening advice when, suddenly, we were all at a loss for where to turn.

At Princeton University Press Brigitta van Rheinberg met my project with enthusiasm and provided me with valuable guidance on how I might make it better. Clara Platter gracefully handled my queries, and Will Hively provided very helpful copyediting. I am grateful to Cambridge University Press for allowing me to reprint with permission portions of this book that appeared in 2005 as the article "Stalin as the Coryphaeus of Science," in *Stalin: A New History*, edited by Sarah Davies and James Harris. I would also like to thank the Maxwell School of Citizenship and Public Affairs at Syracuse University for providing timely and generous grants for last-minute research and production costs.

Two people, David Engerman and Stephen Bittner, deserve special thanks for helping me make sense of documents, reading many drafts of every chapter, tirelessly offering suggestions on revisions, and, in their separate ways, keeping me focused and in good spirits while I worked on this book. Finally, I would like to thank my family for providing a home alive with curiosity and debate. My brothers Josh and Aaron eased my way in life and sparked my interest in history. My mom and dad have always been especially supportive. The stability they created at home made it possible for me to explore faraway places with the confidence that I would always be welcomed back. I dedicate this book to them.

As anyone who has made it this far in these acknowledgments can tell, I have had my share of good fortune. Still, my greatest bit of luck came when Amy Mendillo chose the table next to me. I thank her for her patience, humor, insight, and love.

INDEX

Academician Ivan Pavlov (Ministry of Cinematography film), 142–43

Academy of Medical Sciences, 136–67

Academy of Sciences, 5, 17, 23, 43–44, 210; "Against Vulgarization and Perversion of Marxism in Linguistics," 132–33; and Agricultural Academy meeting, 63–68; and linguistics discussion, 110–12, 131; meeting on "species competition," 48–49; Pavlov Institute of Physiology, 159; and Pavlov Session, 136–67; and physics discussion, 73, 75, 83, Scientific Secretariat, 91–92; Section on Literature and Language, 111. *See also* academy physicists

academy physicists: and physics discussion, 97–101; and university physicists, 73–75, 81–86. *See also names of individuals*

Acharian, R. A., 107, 113

Agitprop (Agitation and Propaganda Administration), 8, 23, 39, 46, 55, 82, 117, 160, 176, 207, 226n23; Aleksandrov and, 16, 18; and cancellation of physics conference, 91–92; leadership changes in, 37–38; and Soviet linguistics, 110–14, 127–33. *See also* Science Section; Shepilov, Dmitrii; Suslov, Mikhail; Zhdanov, Yuri

Agricultural Academy. *See* Lenin All-Union Agricultural Academy

agriculture, 41–71, 181. *See also* biology discussion

Airapet'iants, E. Sh., 160

Akulov, N. S., 89, 98

Aleksandrov, Anatolii, 95

Aleksandrov, Georgii, 9; and biology discussion, 58; *History of Western European Philosophy*, 15–40, 75; *Istoriia filosofii* (1943), vol. 3, 17–18; and philosophy discussion, 15–40, 214; and physics discussion, 81; and physiology discussion, 156; and political economy discussion, 171–74, 193

Alikhanian, S. I., 64, 69

All-Union Conference of Physicists, 72–73; cancellation of, 90–93, 102; planning for, 83–90

Alpatov, V. M., 127

Al'tshuler, Lev, 99, 215

Anokhin, Petr Kuzmich, 139, 151–53, 157, 159, 163

anticosmopolitanism, 7, 99–100, 177–79, 214–15. *See also* anti-Semitism

anti-Marrists, 110

anti-Pavlovians, 145, 148, 151, 162–65

anti-Semitism, 81–83, 99–100, 178–79, 214–15. *See also* Jews

Arakcheev regimes, 125–26, 153, 161, 193

Arbatov, Georgii, 219

archival research, 11–14, 170, 227n28, 247n50

Armenian scholars, 113

Artsimovich, L. A., 98

Arutiunov, A. G., 113

Asratian, Ezras Astratovich, 154, 159, 161, 244n99

atomic physicists, 97–102. *See also* physics

atomic project, Soviet, 73, 92–93, 100–102

"atomic shield," 99, 215